Selected Problems of Contemporary Thermomechanics

Edited by Jerzy Winczek

Published in London, United Kingdom

IntechOpen

Supporting open minds since 2005

Selected Problems of Contemporary Thermomechanics
http://dx.doi.org/10.5772/intechopen.72233
Edited by Jerzy Winczek

Contributors
Hilbeth P. Azikri De Deus, Mikhail Itskov, Jeongho Kim, Sushrut Vaidya, Mariusz Banaszkiewicz, Janusz Badur, Jerzy Winczek

First published in London, United Kingdom, 2018 by IntechOpen
IntechOpen is the global imprint of INTECHOPEN LIMITED, registered in England and Wales, registration number: 11086078, The Shard, 25th floor, 32 London Bridge Street
London, SE19SG - United Kingdom
Printed in Croatia

British Library Cataloguing-in-Publication Data
A catalogue record for this book is available from the British Library

Additional hard copies can be obtained from orders@intechopen.com

Selected Problems of Contemporary Thermomechanics
Edited by Jerzy Winczek
p. cm.
Print ISBN 978-1-78923-762-7
Online ISBN 978-1-78923-763-4

Meet the editor

Jerzy Winczek is professor of Czestochowa University of Technology in Poland. His scientific interest area includes mechanics and thermomechanical states (temperature field, phase changes, strains and stresses, modelling, analytical and numerical methods) in technological processes of metals and their alloys. He received his MSc Eng in 1983, PhD in 1994 and DSc in 2013. Prof. Winczek is the author of monographs and over 100 publications in journals, conference proceedings and chapters in books. He is a member of numerous scientific societies, editorial boards of journals and scientific committees of conferences. Prof. Winczek participated in many research projects funded from national central programs. For his achievements, he was awarded Silver Cross of Merit by President of the Republic of Poland (2016).

Contents

Chapter 1

Introductory Chapter: Introduction to Thermomechanics Problems

Jerzy Winczek

1. Introduction

Thermomechanics is a scientific discipline which investigates the behavior of bodies (solid, liquid, and gas) under the action forces and heat input. Thermomechanical phenomena commonly occur in the human environment, from the action of solar radiation to the technological processes. The analysis of these phenomena often requires extensive interdisciplinary knowledge, e.g., thermodynamics, continuum mechanics (solid and fluid), soil mechanics, biomechanics, metallurgy, hydraulics, civil engineering, and materials science and even anatomy, chemistry, meteorology, or hydrology. The wide range of thermomechanics applications depends on the field of science and the areas of knowledge in which phenomena are considered. The description of these phenomena requires not only knowledge of the laws of physics but the use of advanced mathematical apparatus, tensor algebra, and methods for solving differential and integral equations. Thermomechanical phenomena are analyzed using analytical and numerical methods. The analytical solution offers a quicker assessment of the searched values and its dependence on the various parameters, but for more complex problems, they are difficult or even impossible to apply. Some problems can be solved only with numerical methods, of which the finite element method is commonly used, but also methods of boundary elements, finite differences and elementary balances. In addition to the mentioned above methods, one needs to know how to solve complex equation systems (in case of the author's original software) or to possess the ability to handle professional engineer's packages.

Thermomechanics therefore describes a broad category of phenomena. It is a generalization of classical mechanical theory and thermodynamic theory. Currently, thermomechanical coupling is a fully formed issue. Basic dependences and differential equations have been formulated based on mechanical and thermodynamic laws. Numerous methods and algorithms for solving differential equations of thermomechanical coupling have been developed, including the finite element method.

Looking at the development of thermomechanics, we cannot omit scientists who laid the foundation for this area of science. First and foremost, Isaak Newton, the author of the three principles of dynamics [1], an outstanding physicist and mathematician, parallels with G.W. Leibniz who developed the theory of differential and integral calculus. In turn, the development of thermomechanics (and not only) was contributed by Fourier, the creator of the Fourier transform and Fourier series theories, which he used in his fundamental work on the theory of heat conduction [2]. One should also mention eminent scientists, the creator of the law of thermal radiation, Kirchhoff [3] and Maxwell [4, 5]. Over the past half-century, a number of books have been published that take into account the mutual coupling of thermal and mechanical phenomena, among which one can mention, for example, books written by Gibbings [6], Wilmański [7], Mićunović [8], Ziegler [9], Hsu [10], Maugin [11], Bermudes [12], Nicholson [13], Jou et al. [14], Consiglieri [15], or Kleiber and Kowalczyk [16].

Modern solutions of thermomechanics concern mainly the solid body [7–9, 11–13, 16] and fluids [6, 9, 14, 15], as inanimate nature objects. Increasingly, however, the interests of scientists turn to the analysis of the human body [17, 18]. The scale of the considered phenomena is also widening from the macro scale, through micro, to nano [19, 20].

Because the whole world around us is subject to the laws of mechanics (everything that moves under the laws of kinematics or dynamics), and at the same time constantly real objects are subject to the influence of heat, thermomechanics becomes ubiquitous in human life.

Despite extensive literature, the number of practical examples of using thermomechanics to solve engineering problems is still insufficient.

This book intends to present current trends and methods in solving thermomechanical problems.

Author details

Jerzy Winczek
Faculty of Mechanical Engineering and Computer Science, Czestochowa University of Technology, Czestochowa, Poland

*Address all correspondence to: winczek@imipkm.pcz.pl

References

[1] Newton I. Philosophiae Naturalis Principia Mathematica. London: Juffu Societatis Regre ac Typis Josephi Streater Proftant Venales apud Sam. Smith ad infignia Principis Walia in Coemiterio D. Pauli; 1687. 511 p

[2] Fourier JB. Théorie Analytique de la Chaleur. Paris: Firmin Didot; 1822. 639 p

[3] Kirchhoff RG. Vorlesungen über Mathematische Physik: Mechanik. Leipzig: Druck und Verlag von G.G. Teubner; 1876. 466 p

[4] Maxwell JC. Matter and Motion. London: Society for Promoting Christian Knowledge; 1876. 163 p

[5] Maxwell JC. Theory of Heat. London: Longmans; 1871. 348 p

[6] Gibbings JC. An Introduction to the Governing Equations of Thermodynamics and of the Mechanics of Fluids. Oxford, London: Pergamon Press; 1970. 302 p

[7] Wilmański K. Thermomechanics of Continua. Berlin: Springer; 1998. 273 p

[8] Mićunović MV. Thermomechanics of Viscoplasticity. Fundamentals and Applications. New York: Springer; 2009. 249 p

[9] Ziegler H. An Introduction to Thermomechanics. Amsterdam, New York: North-Holland; 1983. 370 p

[10] Hsu T-R. The Finite Element Method in Thermomechanics. Boston: Allen & Unwin; 1986. 391 p

[11] Maugin GA. The Thermomechanics of Plasticity and Fracture. Cambridge University Press; 1992. 350 p

[12] Bermudes AB. Continuum Thermomechanics. Basel: Birkhäuser Verlag; 2000. 209 p

[13] Nicholson DW. Finite Element Analysis: Thermomechanics of Solids. 2nd ed. Boca Raton: CRC Press Taylor & Francis Group; 2008. 457 p

[14] Jou D, Casas-Vázquez J, Criado-Sancho M. Thermodynamics of Fluids Under Flow. Dordrecht, Heidelberg, London, New York: Springer; 2011. 255 p

[15] Consiglieri L. Mathematical Analysis of Selected Problems from Fluid Thermomechanics. Colne: Lambert Academic Publishing; 2011. 188 p

[16] Kleiber M, Kowalczyk P. Introduction to Nonlinear Thermomechanics of Solids. Basel: Springer International Publishing Switzerland; 2016. 345 p

[17] Ciesielski M, Mochnacki B. Numerical simulation of the heating process in the domain of tissue insulated by protective clothing. Journal of Applied Mathematics and Computational Mechanics. 2014;**13**(4): 91-96

[18] Massoudi M, Kim J, Antaki JF. Modeling and numerical simulation of blood flow using the theory of interacting continua. International Journal of Non-Linear Mechanics. 2012; **47**(5):506-520

[19] Hill TL. Extension of nanothermodynamics to include a one-dimensional surface excess. Nano Letters. 2001;**1**(3):159

[20] Qian H. Hill's small systems nanothermodynamics: A simple macromolecular partition problem with a statistical perspective. Journal of Biological Physics. 2012;**38**(2):201-207

Chapter 2

Thermomechanics of Solid Oxide Fuel Cell Electrode Microstructures Using Finite Element Methods: Progressive Interface Degradation under Thermal Cycling

Sushrut Vaidya and Jeong ho Kim

Abstract

The electrochemical performance of solid oxide fuel cell (SOFC) is significantly influenced by three-phase boundary (TPB) zones in the microstructure. TPB zones are locations where all three phases comprising the microstructure such as the two solid phases and the pore phase are present. Electrochemical reactions such as oxygen reduction occur near TPBs, and TPB density is believed to affect the polarization resistance of the cathode. In this regard, the effect of interface degradation under repeated thermal loading on the mechanical integrity and electrochemical performance of solid oxide fuel cell (SOFC) electrodes is studied through finite element simulations. Image-based 3-D models are used in this study, with additional interface zones at the boundaries between dissimilar solid phases. These interface zones are composed of 3-D cohesive elements of small thickness. The effect of interface degradation on mechanical integrity is studied by subjecting 50:50 LSM:YSZ wt.% cathode models to increasing levels of thermal load from room temperature (20°C) up to operating temperature (820°C). Energy quantities (e.g., strain energy and damage dissipation) for cathode models with and without cohesive interface zones are obtained through finite element analysis (FEA). These quantities are compared using energy balance concepts from fracture mechanics to gain insight into the effects of interface degradation on mechanical integrity.

Keywords: solid oxide fuel cells, cathode, thermal cycling, probability of failure, finite element analysis

1. Introduction

Solid oxide fuel cells (SOFCs) are capable of converting chemical energy into electrical energy with high efficiency and low emissions [1]. Anode, cathode, electrolyte, and interconnect wires are the basic components of SOFCs [1]. SOFCs have been investigated for such configurations [2, 3], materials [4–8], component microstructures [9–12], electrochemical performance [11–14], and thermal stresses [14–17]. A variety of approaches have been chosen to investigate problems related to SOFCs: analytical [3], experimental [4–12], and computational [2, 13–17]. Research has shown that electrochemical performance of SOFCs is affected by component microstructure [11, 12]. In contrast, few studies have investigated the effects of microstructure on mechanical performance [16]. Mechanical durability of the SOFC is a significant factor for performance. It is important to understand the effect of microstructure on durability and probability of failure under thermal loads.

We perform finite element analysis (FEA) of thermal stresses induced in reconstructed SOFC cathode microstructures under thermal cycling. 3-D finite element (FE) models of SOFC cathode microstructures are generated from a stack of 2-D microstructure images. Anode (50:50 wt.% NiO:YSZ) and cathode (50:50 wt.% LSM:YSZ) microstructures have previously been analyzed and validated for thermal stress using finite elements by the authors [17]. This study extends the work by investigating the effect of interface degradation under repeated thermal loading on the mechanical integrity and electrochemical performance of SOFC electrodes through finite element simulations. The effect of interface degradation on mechanical integrity is studied by subjecting 50:50 LSM:YSZ wt.% cathode models to increasing levels of thermal load from room temperature (20°C) up to operating temperatures (820°C). Energy quantities (e.g., strain energy and damage dissipation) for cathode models with and without cohesive interface zones are obtained through FEA. These quantities are compared using energy balance concepts from fracture mechanics to gain insight into the effects of interface degradation on mechanical integrity. The electrochemical performance of SOFCs is significantly influenced by three-phase boundary (TPB) zones in the microstructure. TPB zones are locations where all three phases comprising the microstructure—the two solid phases and the pore phase—are present. Electrochemical reactions such as oxygen reduction occur near TPBs, and TPB density is believed to affect the polarization resistance of the cathode [5, 18]. In this study, we hypothesize that degradation of weak interfaces under thermal cycling has an adverse effect on TPB zones in the microstructure of the SOFC, leading to a reduction in electrochemical performance over time. Interface degradation under thermal cycling is implemented in the FE electrode models through a simplified scheme. The scheme consists of five successive monotonic, steady-state heating operations from room temperature (20°C) up to operating temperature (820°C), combined with interface strength and fracture energy degradation in each heating operation. Each thermal loading operation represents the heating phase of a normalized thermal cycle, where one normalized cycle represents 1000 actual thermal cycles. SOFCs have been known to perform with great reliability, showing no reduction in performance, for more than 2 years at a stretch [1]. Thus, it may be reasonable to assume that one normalized heating cycle in this study represents 1000 actual thermal cycles in the operating life of the cell. The interface degradation scheme is explained in detail later.

2. Image-based finite element microstructure models

3-D FE microstructure models were reconstructed from 41 2-D cross-sectional images of cathode microstructures [9, 10], using focused ion beam-scanning electron microscopy (FIB-SEM). Examples of the 2-D images are shown in **Figure 1**. These images are of the real cathode microstructures having 50:50 LSM:YSZ composition.

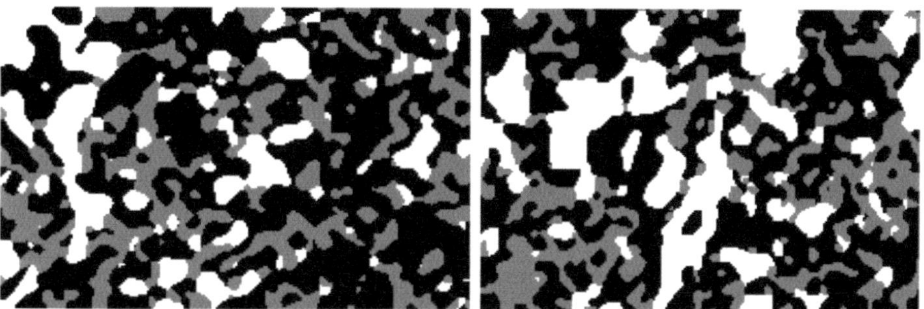

Figure 1.
2-D SEM images of typical cross-sections of 50:50 wt.% LSM:YSZ cathode [9, 10].

In the SEM cathode images, white (pixel value = 255) represents LSM, gray (pixel value = 127) represents YSZ, and black (pixel value = 0) represents the pores. The original cathode image has 217 pixels (width) × 147 pixels (height). The x-y (in-plane) spatial resolution between pixels is 40.8 nm and the z-spacing between images is 53.3 nm. The cathode images are thus of size 8.85 μm (width) × 6.00 μm (height). Finite element modeling was carried out using the commercial FE software Abaqus [19] for which MATLAB® [20] was to create input files. The 3-D FE model was reconstructed from the 2-D images using 3-D finite elements (eight-node brick elements). In order to increase computational efficiency for 3-D analysis using finite element method, we reduced the full model to a representative model in which we sacrifice some details of phase geometry but microstructural skeleton that is crucial to stress analysis remains almost unchanged. To provide validation data for this present approach, we used 2-D cathode microstructure images and calculated the three-phase boundary (TPB) density and phase surface area density for the original, full-size 50:50 wt.% LSM:YSZ cathode, as well as the reduced 50:50 cathode. The data agreed reasonably well with those reported in the literature [10, 12, 21].

The original images of the cathode were simplified by sampling pixels at regular intervals to reduce the image resolution while retaining a detailed microstructure. The simplified cathode images were 3.51 μm (width) × 3.02 μm (height) in size. The depth of the voxel was 53.3 nm. A stack of all the 2-D images was created in the z-direction by using a cell array construct to arrange the images consecutively. An initially blank "buffer" plane was then introduced between consecutive images. The gaps between consecutive images were filled by assigning one eight-node brick element to each voxel. Thus, the geometry of the cathode microstructure was reconstructed in the 3-D FE models, with a step variation in material regions between consecutive images. The volume fractions of phases in each 3-D model were calculated by counting the number of voxels corresponding to each phase (based on pixel value) and dividing by the total number of voxels in the model. Information concerning the material properties, boundary conditions, initial temperature, temperature field and required outputs was also specified in the input file.

3. Thermomechanical material properties

Finite element analyses of the LSM/YSZ (50:50 wt.%) cathode microstructure models were carried out to investigate thermal stresses due to various temperature fields. The effects of varying phase volume fractions and temperature-dependent material properties on thermal stresses and probability of failure were investigated. The FE model was subjected to fixed boundary conditions. The behavior of the

Material	Young's modulus (GPa)	Poisson's ratio	CTE (10^{-6} °C^{-1})
YSZ	205	0.30	10.40
LSM	40	0.25	11.40

Table 1.
Room temperature material properties used in FE analyses of cathode [3, 22].

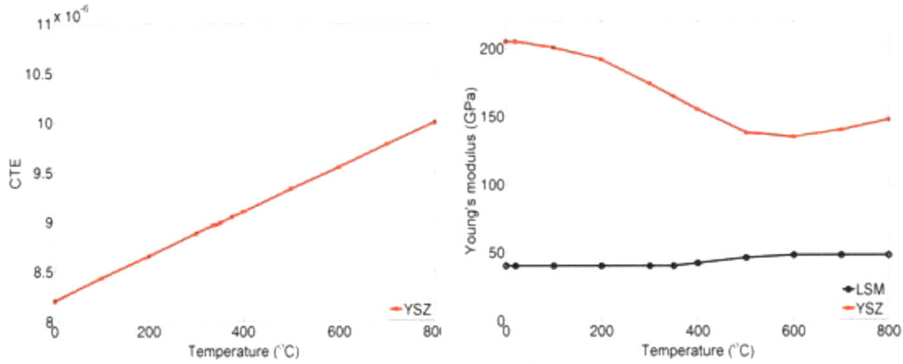

Figure 2.
Variation of CTE of YSZ with temperature [16]; variation of Young's modulus of LSM and YSZ with temperature [12].

model with increasing temperature loads was investigated by subjecting the model to eight different spatially uniform predefined temperature fields of magnitude 120, 220, 320, ..., 820°C. In each analysis, the initial temperature was specified as 20°C (room temperature), so that the model was subjected to eight different magnitudes of temperature change (ΔT = 100, 200, 300, ..., 800°C). **Table 1** lists the room temperature material properties used for YSZ and LSM [3, 22].

Figure 2 shows the variation of the CTE of YSZ with temperature [22] and the variation of Young's modulus of LSM and YSZ with temperature [4]. The CTE of LSM was assumed constant over the temperature range considered. The room temperature value of the CTE of LSM was used in the FE analyses.

4. Finite element analysis using cohesive interface model

An interface damage initiation and evolution model is implemented in the 3-D finite element (FE) SOFC electrode microstructures. The thin interface zones between dissimilar solid phases consist of 3-D cohesive elements. The 3-D cohesive elements, which are of negligible thickness compared with the neighboring solid elements, are used to simulate debonding between different solid phases at their interface. The 3-D cohesive elements are inserted between eight-node 3-D linear brick elements belonging to dissimilar solid phases, as shown in **Figure 3**. Based on the in-plane (x-y) dimensions of the solid elements in the anode (14.0 nm) and cathode (40.8 nm), the thickness of the thin interface cohesive elements is chosen as 1 nm. Interfacial (fracture) energy is used as a reference value for bond strength between LSM and YSZ (cathode). Perfect bonding is assumed between elements belonging to the same solid phase. An algorithm based on the boundary pixel identification scheme is used to identify elements lying on solid-phase boundaries.

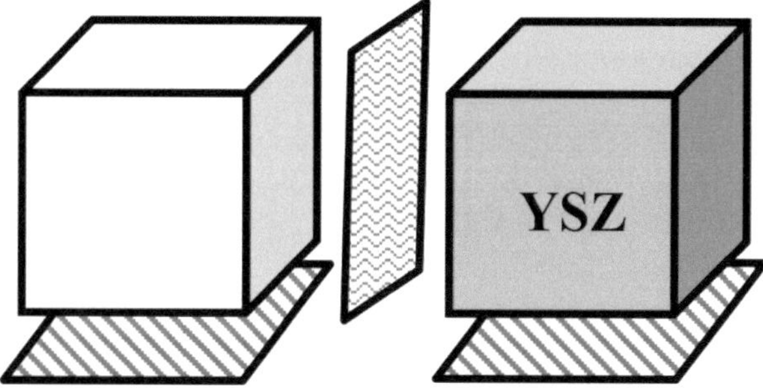

Figure 3.
Cohesive elements in the interface zone between solid elements of different phases. Vertical and horizontal cohesive elements are used.

4.1 Traction-separation behavior of cohesive elements

The behavior of the interface cohesive zone elements is described using a traction-separation relation. Such a cohesive zone model (CZM) describes local material separation behavior by relating tractions ($\boldsymbol{t}$) on the interface surfaces to material separation ($\boldsymbol{\delta}$). The CZM simulates progressive damage in the cohesive zone. The CZM is used to model imperfect bonding in highly porous electrodes of SOFCs. In three dimensions, the traction-separation model for cohesive elements consists of three deformation modes: one normal mode and two in-plane shear modes. In the normal mode, the deformation is purely in a direction perpendicular to the plane of the interface. In the two in-plane shear modes, the deformation is parallel to the plane of the interface. Thus, for 3-D cohesive elements, the traction vector ($\boldsymbol{t}$) has three components: one normal component (i.e., perpendicular to the plane of the interface), t_n, and two in-plane shear components, t_s and t_t, parallel to the plane of the interface. The separation vector ($\boldsymbol{\delta}$) has three corresponding components, δ_n, δ_s, and δ_t. The nominal strain components can then be written as follows:

$$\varepsilon_n = \frac{\delta_n}{T_0}; \varepsilon_s = \frac{\delta_s}{T_0}; \varepsilon_t = \frac{\delta_t}{T_0} \tag{1}$$

Here, T_0 is the initial thickness of the interface. If we set $T_0 = 1$, the nominal strain components become numerically equal to the corresponding separation components. This rescaling can be achieved by defining the interface stiffness (K_i) as follows:

$$K_i = \frac{E_i}{T_0} \tag{2}$$

E_i is the Young's modulus of the interface material and T_0 is the initial thickness of the interface. In this study, the weak interfaces between dissimilar solid phases are modeled using cohesive elements with linear elastic traction-separation behavior up to damage initiation, followed by linear stiffness degradation, to simulate damage evolution. A linear stiffness degradation model has been used by Xiao et al. [23] to study the fracture behavior of ultra-high strength concrete. Thus, the linear damage evolution model may be appropriate for describing the degradation of SOFC electrode interfaces between dissimilar solid (ceramic) phases that exhibit brittle behavior. The normal and in-plane modes of the traction-separation model used in this study are assumed to be uncoupled. The uncoupled linear elastic constitutive behavior of the cohesive elements can be expressed as follows, in terms of tractions $\boldsymbol{t}$ and separations $\boldsymbol{\delta}$ (which are numerically equal to strains ε):

$$\boldsymbol{t} = \begin{Bmatrix} t_n \\ t_s \\ t_t \end{Bmatrix} = \begin{bmatrix} K_{nn} & 0 & 0 \\ 0 & K_{ss} & 0 \\ 0 & 0 & K_{tt} \end{bmatrix} \begin{Bmatrix} \varepsilon_n \\ \varepsilon_s \\ \varepsilon_t \end{Bmatrix} = \boldsymbol{K_i \varepsilon} \tag{3}$$

The uncoupled behavior of the cohesive elements is indicated by the fact that all off-diagonal terms are zero in the interface stiffness ($\boldsymbol{K}_i$) matrix in the constitutive traction-separation ($\boldsymbol{t}$-$\boldsymbol{\delta}$) relation given earlier. **Figure 4** is a schematic representation of the traction-separation curve for LSM/YSZ interface cohesive elements used in this study. In **Figure 4**, δ^0 is the critical separation at which damage is initiated in the cohesive element. Similarly, δ^f is the final separation at the ultimate failure of the cohesive element.

Based on the approach adopted by Nguyen et al. [24], a quadratic stress interaction equation is used to describe the damage initiation condition:

$$\left(\frac{\langle t_n \rangle}{t_n^0}\right)^2 + \left(\frac{t_s}{t_s^0}\right)^2 + \left(\frac{t_t}{t_t^0}\right)^2 = 1 \tag{4}$$

Here, t_n^0, t_s^0, and t_t^0 denote the maximum nominal stress when the deformation is either purely in a direction perpendicular to the plane of the interface or in one of the two orthogonal directions lying in the plane of the interface. The Macaulay brackets < > enclosing t_n indicate that only tensile normal stress causes damage initiation in the cohesive elements. Purely compressive normal stress causes no damage in the cohesive layer [25]. This equation is used to describe the damage initiation criterion for the interface, accounting for the interaction between the normal (tensile) and in-plane (shear) stresses. Due to lack of detailed experimental data on the strength and stiffness properties of the LSM/YSZ interfaces, the approach used by Nguyen et al. [24] is adopted for the cathode: the normal tensile strength and in-plane shear strengths of the LSM/YSZ interface are all assumed to be equal to the Weibull strength of the weaker phase (LSM), that is, 52 MPa [26, 27]. The stiffness of the LSM/YSZ interface is assumed to be characterized by the Young's modulus of

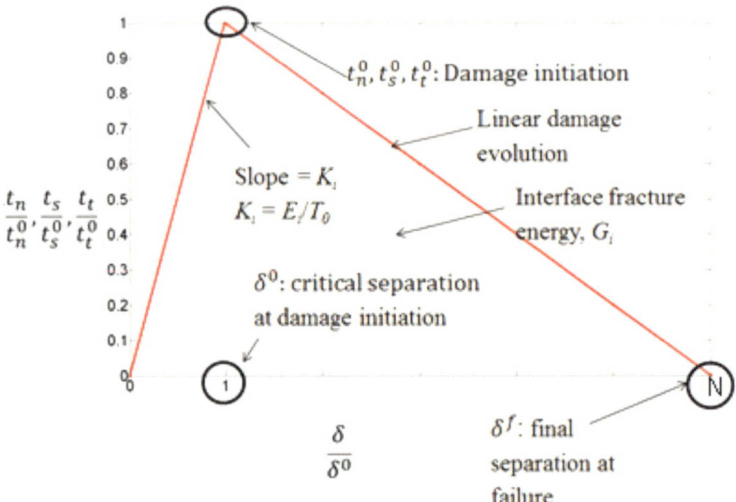

Figure 4.
Schematic diagram of traction-separation curve for an interface.

LSM (40 GPa) [4]. Stiffness degradation does not occur for the cohesive elements under pure normal compressive stress.

The definitions of damage evolution used in this study are based on the fracture energies of the LSM/YSZ interface (cathode). The fracture energy (per unit interface area) is the area under the traction-separation curve for the interface. The critical energy release rate (G_i) for the LSM/YSZ interface is taken as 7.80 J m^{-2} [11]. Based on numerical simulation results reported in the literature [11], the interface is assumed to fail in pure Mode I (normal mode) loading conditions. This leads to a mode-mix ratio ψ = 0° based on traction components. The Young's modulus and original thickness of the LSM/YSZ interface are used to calculate its normalized stiffness K_i (40 x 10^9 N mm^{-3}). The stiffness K_i and interface strength are used to calculate the critical separation δ^0 (0.0013 nm). The area under the traction-separation curve is equal to the fracture energy per unit interface area, 7.80 J m^{-2}. This is used to calculate the final separation at ultimate failure, δ^f (0.3 μm).

4.2 Interface degradation scheme

Degradation of the interface and TPB zones between dissimilar solid phases under repeated thermal cycling may be one of the major reasons for degradation of SOFC performance over time. In this study, interface degradation with thermal cycling is simulated using a simplified scheme. Only the heating phase of the thermal cycle is simulated. Steady-state conditions are assumed and transient effects are neglected. The models are subjected to five successive steady-state heating operations. Each analysis utilizes a spatially uniform temperature field to simulate steady-state heating of the model from room temperature (20°C) up to operating temperature (820°C). Progressive interface degradation is simulated by decreasing the stress at which damage begins in the cohesive interface layers in each successive analysis. The critical and final separations for the cohesive elements are assumed to remain unchanged over the five heating cycles, leading to a progressive reduction of both interface stiffness and critical energy release rate, as shown schematically in **Figure 5**.

Critical energy release rate (i.e., fracture energy per unit interface area) is a measure of the resistance of the interface material to damage [28, 29], while stiffness is indicative of the amount of damage (i.e., stiffness decreases as damage progresses). Therefore, both stiffness and fracture energy are assumed to progressively decrease with thermal cycling. Thus, this scheme implements interface degradation in a

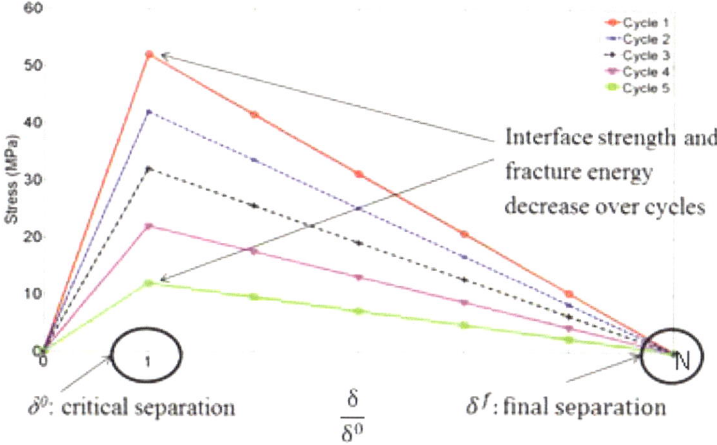

Figure 5.
Schematic diagram of interface degradation model.

simplified manner. Mechanical degradation of the interface is considered, without considering interface degradation due to electrochemical or redox processes. The earlier scheme is only a first approximation to the complex degradation processes that occur at the actual interfaces in a real SOFC electrode. Multiphysics simulations considering both mechanical and electrochemical degradation processes may be an interesting topic for future research.

4.3 Study design

Two sets of electrode microstructure models are generated. One set has perfect bonding throughout the entire model and the other has cohesive zones between dissimilar solid phases. Image-based 3-D models similar to those described in previous sections are used here, with one important difference. The models used are derived through image simplification by analyzing (2×2) pixel squares in the original images. This scheme increases the effective pixel size but preserves both the overall size of the image and the depth of the voxel, leading to a decrease in the reconstructed volume due to the use of a limited number of images. However, this scheme leads to a reasonably good approximation of the original microstructure. In the present study, a similar approach is used to simplify the model. An algorithm is devised to analyze consecutive pixel squares of $(n \times n)$ pixels in each original image. For each such square, the number of pixels belonging to each phase is counted. The phase that contributes the maximum number of pixels to the $n \times n$ pixel square is assigned as the final phase of the single pixel at that location in the simplified image. Additionally, in the present approach, preservation of the original volume of the electrode model in the simplified version is implemented by causing an increase in both effective pixel size and effective voxel depth. The volumes of the simplified models used here are 112.18 μm^3 for the anode and 670.35 μm^3 for the cathode. The cathode volume may be compared against the sampled volume of 178 μm^3 for the simplified cathode model. The values of n used in this study are as follows: for the cathode, $n = 6$, and for the anode, $n = 10$. These levels of simplification are necessary to achieve reasonable computational time for each FE analysis. A detailed validation study for this new approach based on volume fractions, phase surface area densities, and TPB densities may be an interesting topic for future research. In the present study, the plausibility of this new approach is demonstrated by showing that TPB density values obtained for cathode models lie within physically reasonable limits of experimentally determined ranges for these values.

Increasing levels of thermal load are applied through spatially uniform temperature fields of magnitude 120°C, 220°C,..., 820°C, to simulate steady-state heating from room temperature (20°C) up to operating temperature. Energy quantities (e.g., strain energy and damage dissipation) are calculated from FEA for each $\Delta T = 100°C, 200°C,..., 800°C$. The energy quantities for models with and without cohesive zones are compared to gain insight into the effect of weak interface zones between dissimilar solid phases on the mechanical integrity of the cathode. Energy concepts from fracture mechanics are used to interpret the results obtained. This procedure is performed using the cathode microstructure model, considering the original composition (50:50 wt.% LSM:YSZ). In the second part of the study, three-phase boundary (TPB) evolution under repeated, steady-state, monotonic thermal loads is studied using 50:50 wt.% electrode microstructure models. Electrode models with cohesive interface zones are subjected to five successive monotonic, steady-state, heating operations from room temperature up to operating temperature ($\Delta T = +800°C$) using a spatially uniform temperature field. During each successive analysis, the strength of the cohesive interface between dissimilar solid phases is decreased to simulate the effect of interface degradation under thermal cycling. Progressive interface degradation under thermal cycling is implemented using the simplified scheme explained earlier in this chapter. The strains induced in the cohesive layers are calculated using FEA and are used to quantify TPB zone damage using a strain-based criterion, that is, separation in the cohesive layers, as explained later.

4.4 Results and discussion

Cohesive elements give rise to numerical convergence and stability issues that are well documented in the literature. Unstable behavior of cohesive elements due to negative stiffness (i.e., softening) in the damage evolution phase causes local instability in the electrode models used in this study. The local stabilization algorithm available in Abaqus is used along with viscous regularization for the cohesive elements to address convergence and stability issues. Small values of the viscosity parameter (μ = 0.001–0.01) are used to ensure that the energy quantities dissipated in regularization and stabilization are small compared with the strain energy in the model. Fixed boundary conditions (BCs) are prescribed by fixing all degrees of freedom (DOFs) at each node on each face of the models. The 3-D cathode model with cohesive zones is subjected to steady-state temperature change from room temperature (20°C) up to operating temperature. Spatially uniform temperature fields of increasing magnitude (120°C, 220°C,…, 820°C) are used to apply increasing levels of thermal loads (ΔT = 100°C, 200°C,…, 800°C). Similar analyses are performed for the 3-D cathode model without cohesive zones. Energy quantities (e.g., strain energy and damage dissipation) are computed from each FE analysis. **Figure 6** also shows the interface zones between dissimilar solid phases that are composed of cohesive elements.

Figure 7 shows energy quantities obtained from the FE analyses of cathode models with and without cohesive interface zones. The energy quantities are plotted as functions of temperature, which show a comparison between the strain energies of the cohesive model and the non-cohesive model. The energy dissipated due to damage (i.e., the damage dissipation) in the interface zones of the cohesive cathode model is plotted as a function of temperature in **Figure 8**. It is observed from **Figure 7** that the strain energy of the cohesive and non-cohesive cathode models increases with increasing temperature, as expected from the larger thermal stresses induced at higher temperatures, with fixed boundary conditions.

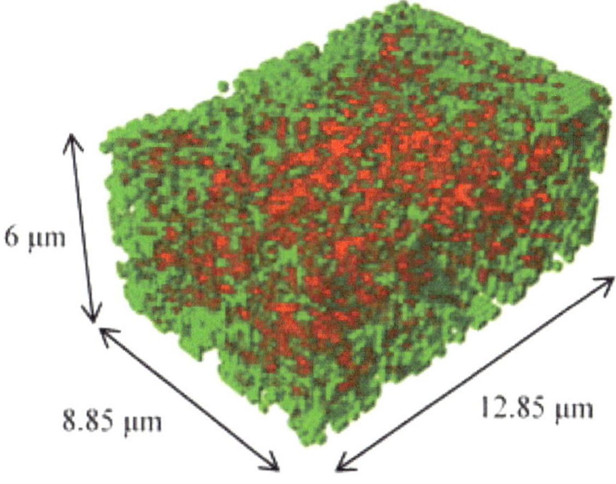

Figure 6.
Cathode model with horizontal interface zones.

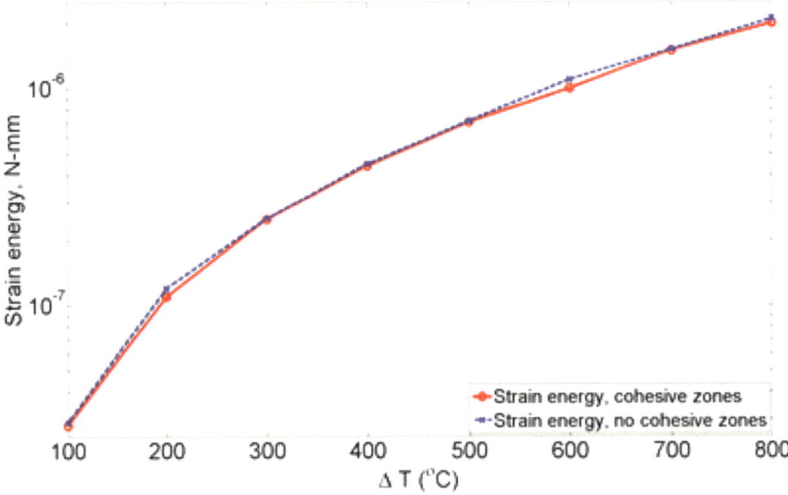

Figure 7.
Strain energy in cathode models with and without cohesive interface zones.

Similarly, from **Figure 8**, it is seen that the energy dissipated due to interfacial damage in the cohesive model also increases with increasing temperature, since more damage occurs at higher temperatures. A significant feature of **Figure 7** is that the strain energy of the non-cohesive model is slightly higher than that of the cohesive model at all temperatures. The physical explanation of the earlier observations is as follows. From the first law of thermodynamics and the Griffith energy balance statement for fracture (i.e., damage), we know that a damage process (e.g., cracking) can occur in a system only if this process causes the total energy of the system to either decrease or remain constant. The limiting condition is attained when damage occurs at equilibrium conditions, with the total energy remaining constant [30]:

$$\frac{dE}{dA} = \frac{d\Pi}{dA} + \frac{dW_s}{dA} = 0 \tag{5}$$

where E is the total energy of the system, Π is the potential energy of the system = U–F, U is the strain energy of the system, F is the work done by external

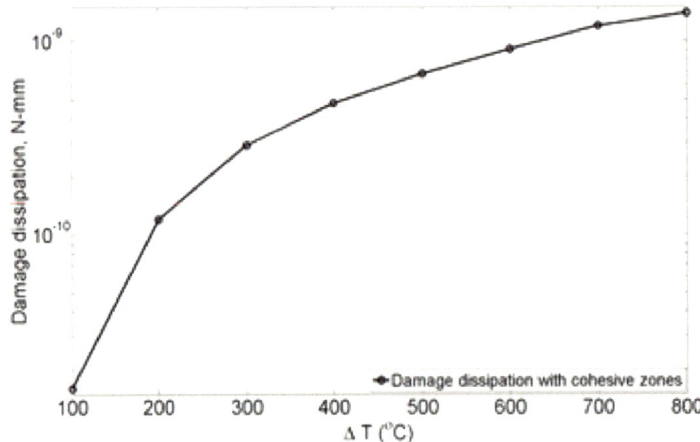

Figure 8.
Damage dissipation in cathode model with cohesive zones.

forces, W_s is the work required for creating new surfaces, and A is the damage variable (e.g., crack length).

In the cathode models analyzed, fixed boundary conditions are prescribed. This means that all translational and rotational DOFs at each node on each face of the model are set to zero. Thus, this model can be regarded as a displacement controlled model [30], with the displacement of the faces fixed at zero. For a displacement controlled system, we know that the external work is zero (F = 0). This is confirmed by the numerical values of the external work obtained from FEA, which are on the order of 10^{-40}. Thus, for a displacement controlled system, Π = U. Finally, from Irwin's energy release rate analysis [30], we know that for displacement control the strain energy of the system decreases as damage (e.g., cracking) occurs. This explains the lower strain energy of the cohesive model as compared to the non-cohesive model at a given temperature.

The second objective of this study is to approximately simulate the effects of thermal cycling on electrode integrity and performance using monotonic, steady-state, thermal loads coupled with an interface strength degradation scheme. In order to achieve this, the electrode models (cathode: 50:50 wt.% LSM:YSZ) with cohesive zones are subjected to five successive analyses. Each analysis represents the heating phase of a normalized thermal cycle, with one normalized cycle being taken equal to 1000 cycles. Only the heating phase of the thermal cycle is simulated in each analysis. Steady-state conditions are assumed and transient effects are neglected. Starting from a uniform initial room temperature of 20°C, the models are subjected to steady-state temperature change up to an operating temperature of 820°C using a spatially uniform temperature field. During each successive analysis, progressive interface degradation is simulated by decreasing the interface strength and fracture energy of the cohesive layers. Two separate interface strength degradation schemes are studied. In the first scheme (Scheme 1), the initial interface strength is assumed to decrease by 5 MPa in each successive cycle. In the second scheme (Scheme 2), the interface strength is assumed to decrease by 10 MPa in each successive cycle. The constitutive thickness of each cohesive element is set equal to 1.0, so that the nominal strain components (ε) calculated using FE analysis are numerically equal to the respective separation components (δ).

Figure 9 shows the interface strength reduction schemes employed to study TPB evolution in the cathode under repeated thermal loading, that is, Scheme 1 ($\Delta t_i^0 = -5$ MPa per cycle, where t_i^0 = interface strength) and Scheme 2 ($\Delta t_i^0 = -10$ MPa per cycle).

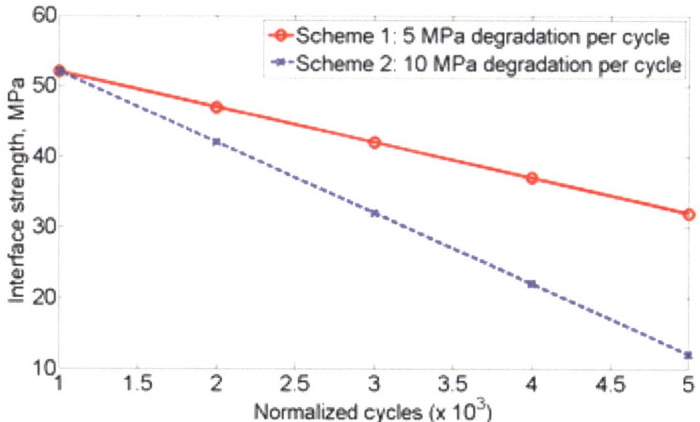

Figure 9.
Interface strength evolution schemes: LSM/YSZ interface.

The numerical values of interface strength and fracture energy used in Scheme 1 (Δt_i^0 = - 5 MPa per cycle) and Scheme 2 (Δt_i^0 = - 10 MPa per cycle) are given in **Table 2**.

Pitakthapanaphong and Busso [11] have performed numerical simulations to determine the critical energy release rates (G_i) for several material interfaces encountered in SOFC systems. Numerically, these values vary from 7.80 J m^{-2} for LSM/YSZ interface to 38.70 J m^{-2} for LSM-LSCoO interface. From these data, it can be concluded that the interface fracture energy values used in this study (**Table 2**) are physically reasonable. Cathode models with fixed boundary conditions (BCs) are used in this study.

The models are subjected to repeated, monotonic, steady-state heating as explained earlier, and the thermal stresses and strains are calculated through FEA. The strains induced in each cohesive element are written to the output data file associated with each analysis, along with strain energy and damage dissipation values. The strains are used to calculate the evolution of three-phase boundaries (TPBs) with thermal cycling, as explained in the next subsection. The energy quantities are used to study the mechanical degradation of the overall model. The TPB length at each temperature is calculated as follows. First, an electrode model without cohesive zones is created by image stacking (using a cell array data structure in MATLAB). The original 2-D images used to reconstruct the 3-D microstructure are stored as pixel value arrays in the cell array. In each 2-D image, boundary pixels are identified by pixel value comparison of each pixel with its eight adjacent in-plane neighbors. Boundary pixels are stored as black pixels (pixel value = 0) at corresponding locations in an initially empty "boundary point" cell array of identical size as the image cell array. Interior (i.e., non-boundary) locations are stored as white pixels at corresponding locations in the boundary point cell array. Using the boundary point cell array, locations of boundary pixels in the image array are identified as black pixels in the boundary point array, and the pixel value of the boundary pixel at the corresponding location in the image array is checked. The pixel values of the eight pixels adjacent to that boundary pixel are also checked to test whether all three phases are present at that boundary location. If so, the location is stored as a three-phase boundary (TPB) point (black pixel) in another initially empty "TPB" cell array. All other non-TPB locations are stored as white pixels in the TPB array. After TPB locations have been identified, the total number of TPB points is counted, and multiplied by the z-distance between images, to obtain the total undamaged TPB length (μm) in the electrode models. The total undamaged TPB length (μm) is divided by the volume (μm^3) of the 3-D electrode model to obtain the TPB density (μm^{-2}) in the model. Damage accumulation over five normalized cycles is calculated by determining the failed TPB cohesive elements in each cycle that had not failed in the previous cycle. The difference between the number of failed TPB elements in the (n + 1)th cycle and in the nth cycle is used to quantify the additional TPB loss in each

Normalized cycles (x 10^3)	**Interface strength, MPa (Scheme 1)**	**Interface strength, MPa (Scheme 2)**	**Interface fracture energy, J m^{-2} (Scheme 1)**	**Interface fracture energy, J m^{-2} (Scheme 2)**
1	52	52	7.80	7.80
2	47	42	7.05	6.30
3	42	32	6.30	4.80
4	37	22	5.55	3.30
5	32	12	4.80	1.80

Table 2.
Cathode model: interface strength and fracture energy data.

successive cycle, and the cumulative TPB loss over N such cycles is thus determined by successively adding the new damage in each cycle to the cumulative total damage over the previous (N - 1) cycles. The model used in this study for evaluating cumulative TPB damage with thermal cycling may be regarded as a simple model for calculating cumulative fatigue damage. A detailed mathematical model for predicting the lifetime of planar SOFCs subjected to thermal cycling has been developed by Liu et al. [31]. That model uses Paris' law and cracks nucleation concepts to derive expressions for damage distribution in the interfacial layers of planar SOFCs under thermal cycling. It predicts that the number of cycles required for failure will decrease with increase in electrolyte thickness and electrode porosity [31].

Figure 10 depicts the change in the strain energy content of the cathode with thermal cycling. **Figure 10** shows that for both interface degradation schemes, the strain energy of the model progressively decreases over five normalized thermal cycles. This is expected from the progressive mechanical degradation of the interfaces within the model with thermal cycling.

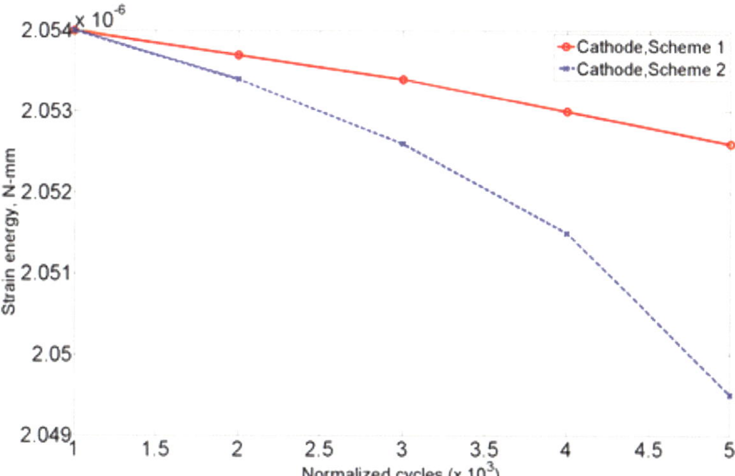

Figure 10.
Strain energy evolution for cathode model with thermal cycling.

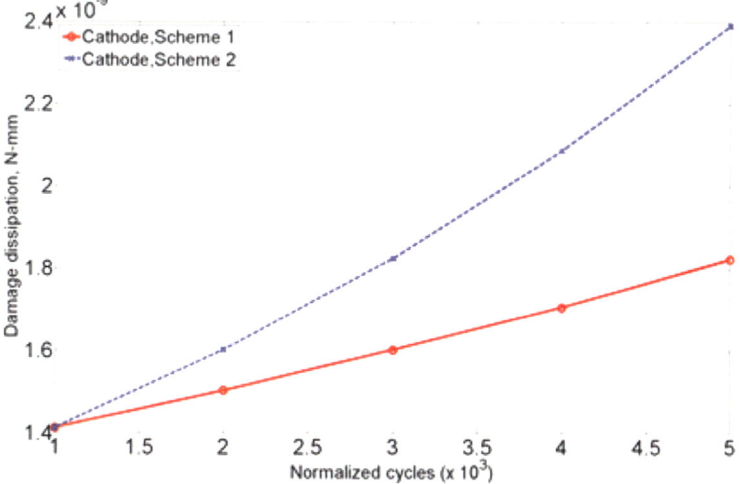

Figure 11.
Damage dissipation in cathode model with thermal cycling.

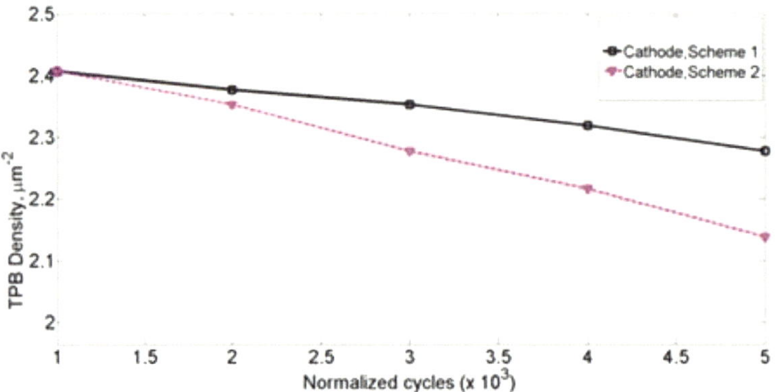

Figure 12.
TPB density evolution with thermal cycling for electrode models.

Such progressive interface degradation leads to a cumulative dissipation of energy due to damage and hence to a progressive decrease in strain energy content of the model. This is further illustrated in **Figure 11**, which shows the cumulative damage dissipation with thermal cycling. As expected, Scheme 2 ($\Delta\, t_i^0 = -10$ MPa per cycle) leads to lower strain energy content, and larger cumulative damage dissipation, than Scheme 1 ($\Delta\, t_i^0 = -5$ MPa per cycle).

Figure 12 shows the evolution of the TPB density with thermal cycling for the anode and cathode models considered in this study. **Figure 12** simultaneously compares the TPB density evolution for cathode versus anode and for Scheme 1 ($\Delta t_i^0 = -5$ MPa per cycle) versus Scheme 2 ($\Delta t_i^0 = -10$ MPa per cycle).

The TPB density of cathode model decreases over the five successive normalized thermal cycles ($\Delta T = +800°C$), as expected from the progressive degradation of the interface with thermal cycling. As expected for the cathode model, Scheme 2 (10 MPa degradation per cycle) leads to a lower final TPB density than Scheme 1 (5 MPa degradation per cycle), after five normalized heating cycles. The TPB density of the cathode undergoes a large reduction (5.36% for Schemes 1, 11.18% for Scheme 2). Wilson et al. [10] have cited TPB density values lying in the range of 1.7–6.5 μm^{-2} for cathodes and have reported cathode TPB density values in the range 6–9 μm^{-2} [12]. From these experimental data, it may be concluded that the TPB density values obtained in this study for the cathode (2.138–2.407 μm^{-2}) are physically reasonable.

5. Conclusion

2-D images of solid oxide fuel cell (SOFC) cathode microstructures (50:50 wt.% LSM:YSZ) are used to perform 3-D finite element models. The effect of LSM/YSZ interface degradation under repeated thermal loading on mechanical integrity and electrochemical performance of SOFC electrodes is simulated by implementing a simplified damage scheme in cathode FE models with cohesive interface zones. The cathode model is first subjected to increasing levels of thermal load using spatially uniform temperature fields. Energy quantities for models with and without cohesive interface zones are obtained through FEA. These quantities are compared using energy balance concepts from fracture mechanics to gain insight into the effects of interface degradation on mechanical integrity. The effect of interface degradation with increasing temperature on the mechanical integrity of a displacement controlled cathode microstructure model is clearly seen as a reduction in strain energy of the model due to damage dissipation. The evolution of three-phase boundary (TPB) zones in electrode microstructure models with thermal cycling [32] is studied by implementing an interface damage scheme that includes reduction of both interface strength and fracture energy. It is found that TPB density decreases over a number of normalized heating cycles. Degradation of the mechanical integrity of the cathode model under repeated thermal loading is also observed, in the form of progressively decreasing strain energy content due to cumulative damage dissipation with thermal cycling. These observations indicate that interface damage may be a major mechanism responsible for SOFC performance degradation over time.

Acknowledgements

The financial support for this work from the National Science Foundation under the Faculty Early Career Development (CAREER) Grant CMMI-0546225 (Material Design & Surface Engineering Program) is gratefully acknowledged.

Conflict of interest

There is no conflict of interest.

Author details

Sushrut Vaidya[1] and Jeong ho Kim[2*]

1 Department of Civil Engineering, Mahindra Ecole Centrale, Hyderabad, Telangana, India

2 Department of Civil and Environmental Engineering, University of Connecticut, Storrs, CT, USA

*Address all correspondence to: jeongho.kim@uconn.edu

References

[1] Singhal SC, Kendall K, editors. High Temperature Solid Oxide Fuel Cells: Fundamentals, Design and Applications. Oxford: Elsevier Ltd.; 2003

[2] Laurencin J, Delette G, Lefebvre-Joud F, Dupeux M. A numerical tool to estimate SOFC mechanical degradation: Case of the planar cell configuration. Journal of the European Ceramic Society. 2008;**28**:1857-1869

[3] Zhang T, Zhu Q, Huang WL, Xie Z, Xin X. Stress field and failure probability analysis for the single cell of planar solid oxide fuel cells. Journal of Power Sources. 2008;**182**:540-545

[4] Giraud S, Canel J. Young's modulus of some SOFCs materials as a function of temperature. Journal of the European Ceramic Society. 2008;**28**:77-83

[5] Selcuk A, Atkinson A. Strength and toughness of tape-cast yttria-stabilized zirconia. Journal of the American Ceramic Society. 2000;**83**(8):2029-2035

[6] Toftegaard H, Sorensen BF, Linderoth S, Lundberg M, Feih S. Effects of heat-treatments on the mechanical strength of coated YSZ: An experimental assessment. Journal of the American Ceramic Society. 2009;**92**(11):2704-2712

[7] Selcuk A, Atkinson A. Elastic properties of ceramic oxides used in solid oxide fuel cells (SOFC). Journal of the European Ceramic Society. 1997;**17**:1523-1532

[8] Pihlatie M, Kaiser A, Mogensen M. Mechanical properties of NiO/Ni-YSZ composites depending on temperature, porosity and redox cycling. Journal of the European Ceramic Society. 2009;**29**:1657-1664

[9] Wilson JR, Kobsiriphat W, Mendoza R, Chen H-Y, Hiller JM, Miller DJ, Thornton K, Voorhees PW, Adler SB, Barnett SA. Three-dimensional reconstruction of a solid-oxide fuel-cell anode. Nature Materials. 2006;**5**:541-544

[10] Wilson JR, Duong AT, Gameiro M, Chen H-Y, Thornton K, Mumm DR, Barnett SA. Quantitative three-dimensional microstructure of a solid oxide fuel cell cathode. Electrochemistry Communications. 2009;**11**:1052-1056

[11] Wilson JR, Barnett SA. Solid oxide fuel cell Ni-YSZ anodes: Effect of composition on microstructure and performance. Electrochemical and Solid-State Letters. 2008;**11**(10):B181-B185

[12] Wilson JR, Cronin JS, Duong AT, Rukes S, Chen H-Y, Thornton K, Mumm DR, Barnett S. Effect of composition of ($La_{0.8}Sr_{0.2}MnO_3$-Y_2O_3-stabilized ZrO_2) cathodes: Correlating three-dimensional microstructure and polarization resistance. Journal of Power Sources. 2010;**195**:1829-1840

[13] Goldin GM, Zhu H, Kee RJ, Bierschenk D, Barnett SA. Multidimensional flow, thermal, and chemical behavior in solid-oxide fuel cell button cells. Journal of Power Sources. 2009;**187**:123-135

[14] Selimovic A, Kemm M, Torisson T, Assadi M. Steady state and transient thermal stress analysis in planar solid oxide fuel cells. Journal of Power Sources. 2005;**145**:463-469

[15] Anandakumar G, Li N, Verma A, Singh P, Kim J-H. Thermal stress and probability of failure analyses of functionally graded solid oxide fuel cells. Journal of Power Sources. 2010;**195**:6659-6670

[16] Clague R, Shearing PR, Lee PD, Zhang Z, Brett DJL, Marquis AJ, Brandon NP. Stress analysis of solid oxide fuel cell anode microstructure reconstructed from focused ion beam tomography. Journal of Power Sources. 2011;**196**:9018-9021

[17] Vaidya S, Kim J-H. Continuum mechanics of solid oxide fuel cells using three-dimensional reconstructed microstructures. In: Gan YX, editor. Continuum Mechanics – Progress in Fundamentals and Engineering Applications. Rijeka: InTech; 2012. pp. 73-88

[18] Zhu W, Ding D, Xia C. Enhancement in three-phase boundary of SOFC electrodes by an ion impregnation method: A modeling comparison. Electrochemical and Solid-State Letters. 2008;**11**:B83-B86

[19] Abaqus v6.11, Dassault Systemes Simulia Corp., Providence, Rhode Island

[20] MATLAB ®R2010a, the MathWorks, Inc., Natick, Massachusetts

[21] Vivet N, Chupin S, Estrade E, Richard A, Bonnamy S, Rochais D, Bruneton E. Effect of Ni content in SOFC Ni-YSZ cermets: A three-dimensional study by FIB-SEM tomography. Journal of Power Sources. 2011;**196**:9989-9997

[22] Johnson J, Qu J. Effective modulus and coefficient of thermal expansion of Ni-YSZ porous cermets. Journal of Power Sources. 2008;**181**:85-92

[23] Xiao J, Schneider H, Dönnecke C, König G. Wedge splitting test on fracture behaviour of ultra high strength concrete. Construction and Building Materials. 2004;**18**:359-365

[24] Nguyen BN, Koeppel BJ, Ahzi S, Khaleel MA, Singh P. Crack growth in solid oxide fuel cell materials: From discrete to continuum damage modeling. Journal of the American Ceramic Society. 2006;**89**:1358-1368

[25] Kakac S, Pramuanjaroenkij A, Zhou XY. A review of numerical modeling of solid oxide fuel cells. International Journal of Hydrogen Energy. 2007;**32**:761-786

[26] Atkinson A, Selçuk A. Mechanical behavior of ceramic oxygen ion-conducting membranes. Solid State Ionics. 2000;**134**:59-66

[27] Ebrahimi F, Bourne G, Kelly MS, Matthews T. Mechanical properties of nanocrystalline nickel produced by electrodeposition. Nanostructured Materials. 1999;**11**:343-350

[28] Janssen M, Zuidema J, Wanhill RJH. Fracture Mechanics. 2nd ed. VSSD. New York: Spon Press; 2006

[29] Meyers M, Chawla K. Mechanical Behavior of Materials. Upper Saddle River, New Jersey, New York: Prentice Hall, Cambridge University Press; 1999

[30] Anderson TL. Fracture Mechanics: Fundamentals and Applications. 4th ed. CRC press; 2005

[31] Liu L, Kim G-Y, Chandra A. Modeling of thermal stresses and lifetime prediction of planar solid oxide fuel cell under thermal cycling conditions. Journal of Power Sources. 2010;**195**:2310-2318

[32] Vaidya S, Kim J-H. Finite element thermal stress analysis of solid oxide fuel cell cathode microstructures. Journal of Power Sources. 2013;**225**:269-276

Chapter 3

On the Thermodynamic Consistency of a Two Micro-Structured Thixotropic Constitutive Model

Hilbeth P. Azikri de Deus and Mikhail Itskov

Abstract

The time-dependent rheological behavior of the thixotropic fluids is presented in various industrial fields (cosmetics, food, oil, etc.). Usually, a couple of equations define constitutive model for thixotropic substances: a constitutive equation based on linear viscoelastic models and a rate equation (an equation related to the micro-structural evolution of the substance). Many constitutive models do not take into account the micro-structural dependence of the shear modulus and viscosity in the dynamic principles from which are developed. The modified Jeffreys model (considering only one single micro-structure type) does not show this incoherence in its formulation. In this chapter, a constitutive model for thixotropic fluids, based on modified Jeffreys model, is presented with the addition of one more micro-structure type, besides of comments on some possible generalizations. The rheological coherence of this constitutive model and thermodynamic consistency are analyzed too. This model takes into account a simple isothermal laminar shear flows, and the micro-structures dynamics are relate to Brownian motion and de Gennes Reptation model via the Smoluchowski™s coagulation theory.

Keywords: thixotropic fluids, thermodynamic consistency, modified Jeffreys model

1. Introduction

The thixotropic substances are considered structured fluids and have a rheological behavior (time-dependent) guided by their structural nature [1–5]. In terms of models (constitutive systems), this behavior is represented by a couple of time-dependent equations. These equations connect the micro-structural characteristics to the rheological behavior. In many works, this system of equations is based on a qualitative way, without any formal justification on the physical principles. In this chapter, a different approach is presented based on some well-established physical principles (Smoluchowski's coagulation theory [6–10], Brownian motion [6–9] and de Gennes reptation model [8, 11, 12]).

The rheological properties evolution, in many thixotropy models [2, 5, 13–17], is presented, generically, in terms of a couple of equations, as follows:

IntechOpen

$$\tau = \tau(\lambda, \dot{\gamma}), \text{ constitutive relationship based on linear viscoelastic models;} \quad (1)$$

$$\dot{\lambda} = \mathfrak{G}_{\dot{\gamma}}(\lambda, \dot{\gamma}), \text{ relation associated to the micro} - \text{structural evolution,} \quad (2)$$

where τ is the shear stress (from Cauchy stress measure), $\dot{\gamma}$ is the shear rate (infinitesimal strain measure), λ is the structural parameter (a positive scalar quantity associated to the structural level of substance) and $\mathfrak{G}_{\dot{\gamma}}$ is a functional form. In large number of works [5, 18–21], Eq. (1) is represented by a linear viscoelastic model/mechanism (Maxwell, Jeffreys, Kelvin-Voigt, etc.) or combinations of that (in parallel or in series). However, it exists a problem with these approaches, they do not consider, the time/micro-structural dependence of the shear modulus (G) and viscosity (η_μ and η_ν) in physical principles from which the constitutive equation system is obtained. Differently, the approach presented in [8] (the modified Jeffreys model) does not present this mistake, considering time/micro-structural evolution in the constitutive system development, and in this way, showing more coherence to represent thixotropic substances behavior.

Eqs. (1) and (2) connect micro-structural characteristic and viscoelasticity nature of the thixotropy. In the specialized literature, a considerable number of micro-structure types [22–29] can be found, although only one single micro-structure type is considered in the development of the modified Jeffreys model. In this way, it is natural to imagine that it can exist more than one micro-structure type in a same thixotropic substance. In this chapter, a new perspective in terms of constitutive approach for thixotropic substances, with apparent-yield-stress nature and based on the modified Jeffreys model [8], is presented in terms of two different types of micro-structure (i.e., two structural parameters), and some of their properties are investigated in a rheological and thermodynamic sense. In the presented approach, each micro-structure is associated to a specific mechanism considered in the model (Maxwell-like element in parallel with a pure viscous Newtonian-like element), that is, one micro-structure type is associated to the Maxwellian element and another one to the pure viscous element. In this way, it is important to stand out that each of them (mechanisms) can react, under the shear loads, in distinct (but integrated) forms one from the other.

Aiming to extend the modified Jeffreys model for two distinguished micro-structures, in the next sections, the constitutive model is formally presented in terms of equations, the analysis of its thermodynamics consistence is done, points related to the characterization of the transition region are discussed and some illustrative numerical simulations of rheological tests are presented as well.

2. Main ideas on two types of micro-structures approach

The approach proposed in this work is based on a Jeffreys model [8] for the thixotropic system representation. In this case, a sketch of the model (Maxwell element in parallel with viscous element) is presented in **Figure 1**.

It is important to point out the presence of two micro-structure types (λ_μ and λ_ν) that are intimately related to the behavior of the shear modulus $G = G(\lambda_\nu)$, and viscosity coefficients $\eta_\nu = \eta_\nu(\lambda_\nu)$ and $\eta_\mu = \eta_\mu(\lambda_\mu)$. In this way, under a certain shear load (τ), for the Maxwell element ($\cdot_\nu$), one has

$$\tau_\nu = \eta_\nu \dot{\gamma}_{\nu_v}; \quad (3)$$

$$\dot{\tau}_\nu = G\dot{\gamma}_{\nu_e} + \dot{G}\gamma_{\nu_e}, \quad (4)$$

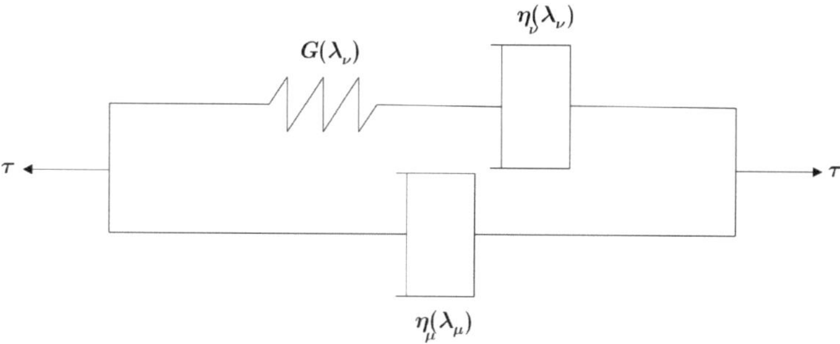

Figure 1.
Sketch of the two micro-structured thixotropic constitutive model.

where $\dot{\gamma}_{\nu_v}$ is the viscous strain rate and $\dot{\gamma}_{\nu_e}$ is the elastic strain rate, respectively, and the total strain rate for the Maxwell element is $\dot{\gamma}_\nu = \dot{\gamma}_{\nu_v} + \dot{\gamma}_{\nu_e}$. After some algebraic manipulation, it follows:

$$\dot{\gamma}_\nu = \frac{\tau_\nu}{\eta_\nu} + \frac{\dot{\tau}_\nu - \dot{G}\gamma_{\nu_e}}{G};$$

$$\therefore \eta_\nu \dot{\gamma}_\nu = \frac{\eta_\nu}{G}\dot{\tau}_\nu + \left(1 - \frac{\eta_\nu \dot{G}}{G^2}\right)\tau_\nu. \tag{5}$$

From the viscous element $(\cdot_\mu)$,

$$\dot{\gamma}_\mu = \frac{\tau_\mu}{\eta_\mu} \therefore \dot{\tau}_\mu = \dot{\eta}_\mu \dot{\gamma}_\mu + \eta_\mu \ddot{\gamma}_\mu, \tag{6}$$

and remembering that $\dot{\gamma} = \dot{\gamma}_\nu = \dot{\gamma}_\mu$ and $\tau = \tau_\nu + \tau_\mu \therefore \dot{\tau} = \dot{\tau}_\nu + \dot{\tau}_\mu$, one has

$$\dot{\tau} = G\dot{\gamma} - \frac{G}{\eta_\nu}\left(1 - \frac{\eta_\nu \dot{G}}{G^2}\right)\tau_\nu + \dot{\eta}_\mu \dot{\gamma} + \eta_\mu \ddot{\gamma}; \tag{7}$$

$$\therefore \frac{\eta_\nu}{G}\dot{\tau} + \left(1 - \frac{\eta_\nu \dot{G}}{G^2}\right)\tau = \left[\eta_\nu + \left(1 - \frac{\eta_\nu \dot{G}}{G^2}\right)\eta_\mu + \frac{\eta_\nu \dot{\eta}_\mu}{G}\right]\dot{\gamma} + \frac{\eta_\nu \eta_\mu}{G}\ddot{\gamma}. \tag{8}$$

It is important to stand out that the time variation of the shear modulus and the viscosity coefficient are taken into account, which is in agreement with the expected physical behavior of thixotropic fluids [8, 30, 31].

It is important to keep in mind the fact that the parameters λ_μ $(0 \le \lambda_\mu \le 1)$ and λ_ν $(0 \le \lambda_\nu \le 1)$ are used to characterize the structural level of the substance. The structural parameter closer to 1 implies highly building-up state of micro-structure, and when it is closer to 0, the micro-structure is close to a fully broken-down state, in respective mechanism $(\cdot_\nu$ or $\cdot_\mu)$. In this sense, it is presented the two micro-structures version of constitutive model based on the work [8].

$$\frac{\eta_\nu}{G}\dot{\tau} + \left(1 - \frac{\eta_\nu \dot{G}}{G^2}\right)\tau = \left[\eta_\nu + \left(1 - \frac{\eta_\nu \dot{G}}{G^2}\right)\eta_\mu + \frac{\eta_\nu \dot{\eta}_\mu}{G}\right]\dot{\gamma} + \frac{\eta_\nu \eta_\mu}{G}\ddot{\gamma}; \tag{9}$$

$$\frac{d\lambda_\nu}{dt} = \frac{1}{t_\nu}\left(k_\nu(1 - \lambda_\nu)^{\beta_\nu}\right) - \frac{(K_\nu^* \lambda_\nu^6 \theta \ddot{\gamma} + \tau_\nu)\lambda_\nu \dot{\gamma}}{\varsigma_\nu}; \tag{10}$$

$$\frac{d\lambda_\mu}{dt} = \frac{1}{t_\mu}\left(k_\mu(1-\lambda_\mu)^{\beta_\mu}\right) - \frac{\left(K_\mu^* \lambda_\mu^6 \theta \ddot{\gamma} + \tau_\mu\right)\lambda_\mu \dot{\gamma}}{\varsigma_\mu}; \tag{11}$$

$$G(\lambda_\nu) = G_o \lambda_\nu^{\aleph} \exp\left(m\lambda_\nu^{-1}\right); \tag{12}$$

$$\eta_\nu(\lambda_\nu) = \eta_o\{\exp\left[(\alpha_1+\alpha_2)\lambda_\nu\right] - \exp\left(\alpha_2\lambda_\nu\right)\}; \tag{13}$$

$$\eta_\mu(\lambda_\mu) = \eta_o \exp\left(\alpha_3\lambda_\mu\right). \tag{14}$$

In this case, there are the following positive parameters: $\aleph$, k_ν, k_μ, β_ν, β_μ, t_ν, t_μ, G_o, m, ς_ν, ς_μ, K_ν^*, K_μ^*, η_o, α_1, α_2 and α_3. Some physical details related to these parameters can be found in [8, 32].

3. On the thermodynamic consistence

In this section, the thermodynamic consistence of constitutive model is analyzed in terms of the entropy-producing processes [33] associated with the concept of natural configuration [8, 34–36]. The main objective is to verify the consistence of the constitutive equation to the Clausius-Duhem inequality.

Figure 2 presents a sketch of the configurations spaces used in this approach for the body $\mathcal{B}$, where $\hat{k}_o(\mathcal{B})$ is the reference configuration, $\hat{k}_t(\mathcal{B})$ is the current/actual configuration, and $\hat{k}_{n(t)}(\mathcal{B})$ is a family of natural configurations. In isothermal process, one is supposed that $\hat{k}_{n(t)}(\mathcal{B}) = \hat{k}_{n(t)}(\mathcal{B})(\lambda_\mu, \lambda_\nu)$, where λ_μ and λ_ν are functions of the flow history. In this sense, it is assumed that the family of natural configurations can be parametrized by the structural level of the substance $(\lambda_\mu, \lambda_\nu)$. In this way, it has $\underline{\underline{F}}_{\hat{k}_o}$ (gradient of the motion from reference configuration to current configuration), $\underline{\underline{F}}_{\hat{k}_{n(t)}}$ (gradient of the motion from natural configuration to current configuration) and $\underline{\underline{F}}_{\hat{k}_o \to \hat{k}_{n(t)}}$ (gradient of the motion from reference configuration to natural configuration). In this way, denoting for each ξ_i-th relaxation mechanism.

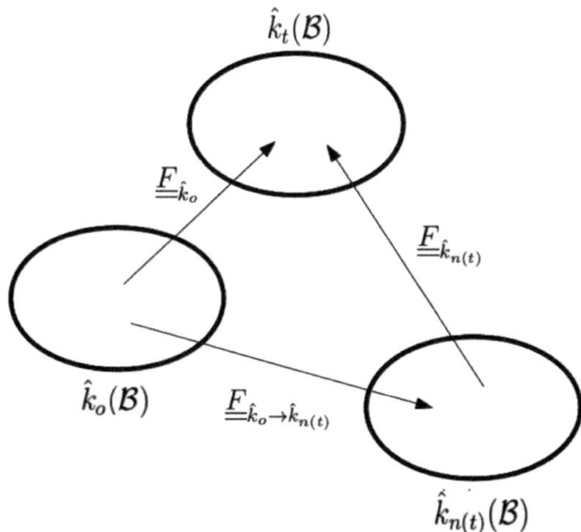

Figure 2.
Configuration spaces.

$$ {}^{\xi_i}\underline{\underline{B}}_{\hat{k}_{n(t)}} = {}^{\xi_i}\underline{\underline{F}}_{\hat{k}_{n(t)}}\,{}^{\xi_i}\underline{\underline{F}}^{T}_{\hat{k}_{n(t)}}; \tag{15} $$

$$ {}^{\xi_i}\underline{\underline{C}}_{\hat{k}_{n(t)}} = {}^{\xi_i}\underline{\underline{F}}^{T}_{\hat{k}_{n(t)}}\,{}^{\xi_i}\underline{\underline{F}}_{\hat{k}_{n(t)}}; \tag{16} $$

$$ {}^{\xi_i}\underline{\underline{L}}_{\hat{k}_{n(t)}} = {}^{\xi_i}\dot{\underline{\underline{F}}}_{\hat{k}_o\to\hat{k}_{n(t)}}\,{}^{\xi_i}\underline{\underline{F}}^{-1}_{\hat{k}_o\to\hat{k}_{n(t)}}; \tag{17} $$

$$ {}^{\xi_i}\underline{\underline{D}}_{\hat{k}_{n(t)}} = \mathrm{sym}\left({}^{\xi_i}\underline{\underline{L}}_{\hat{k}_{n(t)}}\right), \tag{18} $$

and the extensive thermodynamic Helmholtz potential (Ψ) can be represented as

$$ \Psi = \Psi\left(\lambda_\nu, \mathcal{I}_1^\nu, \mathcal{II}_1^\nu, \mathcal{III}_1^\nu, \dots, \mathcal{I}_{r_\nu}^\nu, \mathcal{II}_{r_\nu}^\nu, \mathcal{III}_{r_\nu}^\nu, \lambda_\mu, \mathcal{I}_1^\mu, \mathcal{II}_1^\mu, \mathcal{III}_1^\mu, \dots, \mathcal{I}_{r_\mu}^\mu, \mathcal{II}_{r_\mu}^\mu, \mathcal{III}_{r_\mu}^\mu\right); \tag{19} $$

$$ \therefore \dot{\Psi} = \sum_{i=\nu,\,\mu}\left[\Psi,_{\lambda_i}\dot{\lambda}_i + \sum_{\xi_i=1}^{r_i}\left(\Psi,_{{}^{\xi_i}\underline{\underline{B}}_{\hat{k}_{n(t)}}} : {}^{\xi_i}\dot{\underline{\underline{B}}}_{\hat{k}_{n(t)}}\right)\right], \tag{20} $$

where follows from chain rule $\underline{\underline{\Psi}},_{{}^{\xi_i}\underline{\underline{B}}_{\hat{k}_{n(t)}}} = \left(\Psi,_{\mathcal{I}^i_{\xi_i}} + \mathcal{I}^i_{\xi_i}\Psi,_{\mathcal{II}^i_{\xi_i}}\right)\underline{\underline{I}} - \Psi,_{\mathcal{II}^i_{\xi_i}}\,{}^{\xi_i}\underline{\underline{B}}_{\hat{k}_{n(t)}} + \mathcal{III}^i_{\xi_i}\Psi,_{\mathcal{III}^i_{\xi_i}}\,{}^{\xi_i}\underline{\underline{B}}^{-T}_{\hat{k}_{n(t)}}$, with $\mathcal{I}^i_{\xi_i} = tr\left({}^{\xi_i}\underline{\underline{B}}_{\hat{k}_{n(t)}}\right)$, $\mathcal{II}^i_{\xi_i} = \frac{1}{2}\left[tr\left({}^{\xi_i}\underline{\underline{B}}_{\hat{k}_{n(t)}}\right)^2 - tr\left({}^{\xi_i}\underline{\underline{B}}^2_{\hat{k}_{n(t)}}\right)\right]$ and $\mathcal{III}_{\xi_i} = det\left({}^{\xi_i}\underline{\underline{B}}_{\hat{k}_{n(t)}}\right)$. Without loss of generality it is chosen $\hat{k}_{n(t)}$ such that ${}^{\xi_i}\underline{\underline{F}}_{\hat{k}_{n(t)}} = {}^{\xi_i}\underline{\underline{V}}_{\hat{k}_{n(t)}}$ ($\therefore {}^{\xi_i}\underline{\underline{C}}_{\hat{k}_{n(t)}} = {}^{\xi_i}\underline{\underline{B}}_{\hat{k}_{n(t)}}$), that is, for an appropriate rotated natural configuration $\hat{k}_{n(t)}$, it follows [8]:

$$ \underline{\underline{I}} : {}^{\xi_i}\dot{\underline{\underline{B}}}_{\hat{k}_{n(t)}} = 2\,{}^{\xi_i}\underline{\underline{B}}_{\hat{k}_{n(t)}} : \left(\underline{\underline{D}} - {}^{\xi_i}\underline{\underline{D}}_{\hat{k}_{n(t)}}\right); \tag{21} $$

$$ {}^{\xi_i}\underline{\underline{B}}_{\hat{k}_{n(t)}} : {}^{\xi_i}\dot{\underline{\underline{B}}}_{\hat{k}_{n(t)}} = 2\,{}^{\xi_i}\underline{\underline{B}}^2_{\hat{k}_{n(t)}} : \left(\underline{\underline{D}} - {}^{\xi_i}\underline{\underline{D}}_{\hat{k}_{n(t)}}\right); \tag{22} $$

$$ {}^{\xi_i}\underline{\underline{B}}^{-T}_{\hat{k}_{n(t)}} : {}^{\xi_i}\dot{\underline{\underline{B}}}_{\hat{k}_{n(t)}} = 2\underline{\underline{I}} : \left(\underline{\underline{D}} - {}^{\xi_i}\underline{\underline{D}}_{\hat{k}_{n(t)}}\right), \tag{23} $$

where $\underline{\underline{D}}$ is the symmetric part of velocity gradient ($\underline{\underline{L}}$). Returning to Eq. (20), one has

$$ \begin{aligned} \dot{\Psi} = \sum_{i=\nu,\,\mu}\Bigg\{&\Psi,_{\lambda_i}\dot{\lambda}_i + \sum_{\xi_i=1}^{r_i} 2\left[\left(\Psi,_{\mathcal{I}_{\xi_i}} + \mathcal{I}^i_{\xi_i}\Psi,_{\mathcal{II}^i_{\xi_i}}\right){}^{\xi_i}\underline{\underline{B}}_{\hat{k}_{n(t)}} - \Psi,_{\mathcal{II}^i_{\xi_i}}\,{}^{\xi_i}\underline{\underline{B}}^2_{\hat{k}_{n(t)}}\right. \\ &\left. + \mathcal{III}^i_{\xi_i}\Psi,_{\mathcal{III}^i_{\xi_i}}\underline{\underline{I}}\right] : \left(\underline{\underline{D}} - {}^{\xi_i}\underline{\underline{D}}_{\hat{k}_{n(t)}}\right)\Bigg\}. \end{aligned} \tag{24} $$

Then, from the Clausius-Duhem inequality follows the specific rate of dissipation Ξ

$$\Xi = \underline{\underline{\tau}} : \underline{\underline{D}} - \dot{\Psi}$$

$$= \left\{ \underline{\underline{\tau}} - \sum_{i=\nu,\,\mu} \sum_{\xi_i=1}^{r_i} 2\left[\left(\Psi,_{\mathcal{I}^i_{\xi_i}} + \mathcal{I}^i_{\xi_i} \Psi,_{\mathcal{II}^i_{\xi_i}} \right)^{\xi_i} \underline{\underline{B}}_{\hat{k}_{n(t)}} - \Psi,_{\mathcal{II}^i_{\xi_i}} {}^{\xi_i}\underline{\underline{B}}^2_{\hat{k}_{n(t)}} + \mathcal{III}^i_{\xi_i} \Psi,_{\mathcal{III}^i_{\xi_i}} \underline{\underline{I}} \right] \right\}$$

$$: \underline{\underline{D}} - \sum_{i=\nu,\,\mu} \Psi,_{\lambda_i} \dot{\lambda}_i + \sum_{i=\nu,\,\mu} \sum_{\xi_i=1}^{r_i} 2\left[\left(\Psi,_{\mathcal{I}^i_{\xi_i}} + \mathcal{I}^i_{\xi_i} \Psi,_{\mathcal{II}^i_{\xi_i}} \right)^{\xi_i} \underline{\underline{B}}_{\hat{k}_{n(t)}} - \Psi,_{\mathcal{II}^i_{\xi_i}} {}^{\xi_i}\underline{\underline{B}}^2_{\hat{k}_{n(t)}} \right.$$

$$\left. + \mathcal{III}^i_{\xi_i} \Psi,_{\mathcal{III}^i_{\xi_i}} \underline{\underline{I}} \right] : {}^{\xi_i}\underline{\underline{D}}_{\hat{k}_{n(t)}}, \tag{25}$$

where $\underline{\underline{\tau}}$ is the Cauchy stress tensor.

This approach makes possible the analysis of the constitutive law and the nonnegativity of the entropy production in the context of irreversible thermodynamic processes. In this sense, it is assumed that total rate dissipation Ξ can be divided into three parts, each one associated to a specific process, as follows:

$$\Xi_{\hat{k}_t} = \left\{ \underline{\underline{\tau}} - \sum_{i=\nu,\,\mu} \sum_{\xi_i=1}^{r_i} 2\left[\left(\Psi,_{\mathcal{I}^i_{\xi_i}} + \mathcal{I}^i_{\xi_i} \Psi,_{\mathcal{II}^i_{\xi_i}} \right)^{\xi_i} \underline{\underline{B}}_{\hat{k}_{n(t)}} - \Psi,_{\mathcal{II}^i_{\xi_i}} {}^{\xi_i}\underline{\underline{B}}^2_{\hat{k}_{n(t)}} + \mathcal{III}^i_{\xi_i} \Psi,_{\mathcal{III}^i_{\xi_i}} \underline{\underline{I}} \right] \right\}$$

$$: \underline{\underline{D}} \geq 0; \tag{26}$$

$$\Xi_\lambda = - \sum_{i=\nu,\,\mu} \Psi,_{\lambda_i} \dot{\lambda}_i \geq 0; \tag{27}$$

$$\Xi_{\hat{k}_{n(t)}} = \sum_{i=\nu,\,\mu} \sum_{\xi_i=1}^{r_i} 2\left[\left(\Psi,_{\mathcal{I}^i_{\xi_i}} + \mathcal{I}^i_{\xi_i} \Psi,_{\mathcal{II}^i_{\xi_i}} \right)^{\xi_i} \underline{\underline{B}}_{\hat{k}_{n(t)}} - \Psi,_{\mathcal{II}^i_{\xi_i}} {}^{\xi_i}\underline{\underline{B}}_{\hat{k}_{n(t)}}{}^2 + \mathcal{III}^i_{\xi_i} \Psi,_{\mathcal{III}^i_{\xi_i}} \underline{\underline{I}} \right] : {}^{\xi_i}\underline{\underline{D}}_{\hat{k}_{n(t)}} \geq 0, \tag{28}$$

where $\Xi_{\hat{k}_t}$ is the dissipation rate due to changes in $\hat{k}_t(\mathcal{B})$), Ξ_λ is the dissipation rate due to changes in micro-structure and $\Xi_{\hat{k}_{n(t)}}$ is the dissipation rate due to changes in $\hat{k}_{n(t)}(\mathcal{B})$.

Following similar steps to the work [8], one has

$$\underline{\underline{\tau}} = \sum_{i=\nu,\,\mu} \Sigma_{\xi_i=1}^{r_i} {}^{\xi_i}\underline{\underline{\tau}}, \tag{29}$$

where

$${}^{\xi_i}\underline{\underline{\tau}} = \sum_{i=\nu,\,\mu} \sum_{\xi_i=1}^{r_i} 2\left[\left(\Psi,_{\mathcal{I}^i_{\xi_i}} + \mathcal{I}^i_{\xi_i} \Psi,_{\mathcal{II}^i_{\xi_i}} \right)^{\xi_i} \underline{\underline{B}}_{\hat{k}_{n(t)}} - \Psi,_{\mathcal{II}^i_{\xi_i}} {}^{\xi_i}\underline{\underline{B}}^2_{\hat{k}_{n(t)}} + \mathcal{III}^i_{\xi_i} \Psi,_{\mathcal{III}^i_{\xi_i}} \underline{\underline{I}} \right]. \tag{30}$$

Thus, one has for each ξ_i-relaxation mechanism.

$$\frac{\eta'_{\xi_i}}{G'_{\xi_i}} {}^{\xi_i}\overset{\odot}{\underline{\underline{\tau}}} + \left[1 - \frac{\eta'_{\xi_i} \dot{G}'_{\xi_i}}{G'^2_{\xi_i}} \right]^{\xi_i} \underline{\underline{\tau}} = \eta'_{\xi_i} \underline{\underline{D}}, \tag{31}$$

where $\overset{\odot}{()} = \dot{()} - \underline{\underline{L}}() - ()\underline{\underline{L}}^T$ is the "Oldroyd time derivative". Considering the two relaxation mechanisms, with $r_\nu = r_\mu = 1 \therefore {}^{\xi_\nu}\underline{\underline{\tau}} = \underline{\underline{\tau}}_\nu$ and ${}^{\xi_\mu}\underline{\underline{\tau}} = \underline{\underline{\tau}}_\mu$, where $\eta'_{\xi_\nu} = \eta_\nu$, $G'_{\xi_\nu} = G$, $\eta'_{\xi_\mu} = \eta_\mu$ and $G'_{\xi_\mu} \to +\infty$, one has.

$$\frac{\eta_\nu}{G}\overset{\odot}{\underline{\underline{\tau}}}_\nu + \left[1-\frac{\eta_\nu \dot{G}}{G^2}\right]\underline{\underline{\tau}}_\nu = \eta_\nu \underline{\underline{D}}; \tag{32}$$

$$\underline{\underline{\tau}}_\mu = \eta_\mu \underline{\underline{D}}. \tag{33}$$

Therefore, as $\underline{\underline{\tau}} = \underline{\underline{\tau}}_\nu + \underline{\underline{\tau}}_\mu$ one has

$$\frac{\eta_\nu}{G}\overset{\odot}{\underline{\underline{\tau}}} + \left(1-\frac{\eta_\nu \dot{G}}{G^2}\right)\underline{\underline{\tau}} = \left[\eta_\nu + \left(1-\frac{\eta_\nu \dot{G}}{G^2}\right)\eta_\mu + \frac{\eta_\nu \dot{\eta}_\mu}{G}\right]\underline{\underline{D}} + \frac{\eta_\nu \eta_\mu}{G}\overset{\odot}{\underline{\underline{D}}}. \tag{34}$$

Note that the isochoric motion constraint was not required in this analysis. In this way, associating each relaxation mechanism to a micro-structure, it is important to point out that depending on the nature (level of complexity) of the analyzed thixotropic substance, the reasoning presented here can be extrapolated for more than two relaxation mechanisms (i.e., more than two micro-structure types) aiming a consistent/coherent (in thermodynamic and rheological senses) approach of the model.

4. On the transition region

The objective of this topic is to analyze the transition/yielding region criteria under a theoretical and formal point of view. In this sense, it is important to stand out that the constitutive equation

$$\frac{\eta_\nu}{G}\overset{\odot}{\underline{\underline{\tau}}} + \left(1-\frac{\eta_\nu \dot{G}}{G^2}\right)\underline{\underline{\tau}} = \left[\eta_\nu + \left(1-\frac{\eta_\nu \dot{G}}{G^2}\right)\eta_\mu + \frac{\eta_\nu \dot{\eta}_\mu}{G}\right]\underline{\underline{D}} + \frac{\eta_\nu \eta_\mu}{G}\overset{\odot}{\underline{\underline{D}}}, \tag{35}$$

can be rewritten in a more appropriate form as follows:

$$\theta_1 \overset{\odot}{\underline{\underline{\Pi}}} + \underline{\underline{\Pi}} = \eta\left(\dot{\underline{\underline{\Gamma}}} + \theta_2 \overset{\odot}{\dot{\underline{\underline{\Gamma}}}}\right), \tag{36}$$

where

$$\underline{\underline{\Pi}} = \frac{\underline{\underline{\tau}}}{G}; \tag{37}$$

$$\dot{\underline{\underline{\Gamma}}} = \frac{\eta_\mu}{G}\underline{\underline{D}}; \tag{38}$$

$$\eta = \frac{\eta_\nu + \eta_\mu}{\eta_\mu}; \tag{39}$$

$$\theta_1 = \frac{\eta_\nu}{G}; \tag{40}$$

$$\theta_2 = \frac{\eta_\nu \eta_\mu}{G(\eta_\nu + \eta_\mu)}. \tag{41}$$

It can be seen that the form of Eq. (36) is quite similar to the standard Jeffreys model. The quantities Π, $\dot{\Gamma}$, η, θ_1 and θ_2 can be interpreted as a dimensionless stress, dimensionless strain, dimensionless apparent viscosity, relaxation time and retardation time, respectively. Note that since

$$\dot{\overline{G^{-1}}} = \left(G^{-1}\right)_{,\lambda_\nu}\dot{\lambda}_\nu \to 0;$$

$$\dot{\overline{\frac{\eta_\mu}{G}}} = \left(\frac{\eta_\mu}{G}\right)_{,\lambda_\mu}\dot{\lambda}_\mu + \left(\frac{\eta_\mu}{G}\right)_{,\lambda_\nu}\dot{\lambda}_\nu \to 0,$$

Eq. (36) corresponds to the standard Jeffreys model. In this way, it is clear the relation between the nonlinear viscoelastic region (associated to the thixotropic effect) and $\dot{\lambda}$. Many works propose that an indicative for the beginning of the transition/yielding region is a region where occurs the modification in the behavior of the thixotropic substance, from the linear viscoelastic region to nonlinear viscoelastic region [37–41] associating an specific strain value for that. However, this specific value is not constant [41–45].

In this sense, taking into account the mapping in terms of variables in the actual configurations ($\underline{x} = x^i \underline{g}_i$) and the respective displacements ($\underline{u} = u^i \underline{g}_i$), it follows:

$$\begin{bmatrix} \frac{\partial(x^1-u^1)}{\partial x^1} & \frac{\partial(x^2-u^2)}{\partial x^1} & \frac{\partial(x^3-u^3)}{\partial x^1} \\ \frac{\partial(x^1-u^1)}{\partial x^2} & \frac{\partial(x^2-u^2)}{\partial x^2} & \frac{\partial(x^3-u^3)}{\partial x^2} \\ \frac{\partial(x^1-u^1)}{\partial x^3} & \frac{\partial(x^2-u^2)}{\partial x^3} & \frac{\partial(x^3-u^3)}{\partial x^3} \end{bmatrix} = \begin{bmatrix} 1-\frac{\partial u^1}{\partial x^1} & -\frac{\partial u^2}{\partial x^1} & -\frac{\partial u^3}{\partial x^1} \\ -\frac{\partial u^1}{\partial x^2} & 1-\frac{\partial u^2}{\partial x^2} & -\frac{\partial u^3}{\partial x^2} \\ -\frac{\partial u^1}{\partial x^3} & -\frac{\partial u^2}{\partial x^3} & 1-\frac{\partial u^3}{\partial x^3} \end{bmatrix},$$

and in this context, one has the Jacobian ($\tilde{J}$)

$$\tilde{J} = \det\begin{pmatrix} 1-\frac{\partial u^1}{\partial x^1} & -\frac{\partial u^2}{\partial x^1} & -\frac{\partial u^3}{\partial x^1} \\ -\frac{\partial u^1}{\partial x^2} & 1-\frac{\partial u^2}{\partial x^2} & -\frac{\partial u^3}{\partial x^2} \\ -\frac{\partial u^1}{\partial x^3} & -\frac{\partial u^2}{\partial x^3} & 1-\frac{\partial u^3}{\partial x^3} \end{pmatrix}, \tag{42}$$

and

$$\begin{aligned} \tilde{J}^2 &= \det\begin{pmatrix} 1-2\varepsilon_{11} & -2\varepsilon_{12} & -2\varepsilon_{13} \\ -2\varepsilon_{21} & 1-2\varepsilon_{22} & -2\varepsilon_{23} \\ -2\varepsilon_{31} & -2\varepsilon_{32} & 1-2\varepsilon_{33} \end{pmatrix}; \\ &= 8(\varepsilon_{11}\varepsilon_{23}\varepsilon_{32} + \varepsilon_{12}\varepsilon_{21}\varepsilon_{33} + \varepsilon_{13}\varepsilon_{22}\varepsilon_{31} - \varepsilon_{11}\varepsilon_{22}\varepsilon_{33} - \varepsilon_{12}\varepsilon_{23}\varepsilon_{31} - \varepsilon_{13}\varepsilon_{32}\varepsilon_{21}) \\ &\quad +4(\varepsilon_{11}\varepsilon_{33} + \varepsilon_{11}\varepsilon_{22} + \varepsilon_{22}\varepsilon_{33} - \varepsilon_{12}\varepsilon_{21} - \varepsilon_{13}\varepsilon_{31} - \varepsilon_{23}\varepsilon_{32}) \\ &\quad -2(\varepsilon_{11} + \varepsilon_{22} + \varepsilon_{33}) + 1; \\ &= -8\mathcal{JJJ} + 4\mathcal{JJ} - 2\mathcal{J} + 1, \end{aligned} \tag{43}$$

with

$$2\varepsilon_{ij} = \begin{cases} 2\frac{\partial u^j}{\partial x^i} - \left(\frac{\partial u^1}{\partial x^i}\right)^2 - \left(\frac{\partial u^2}{\partial x^i}\right)^2 - \left(\frac{\partial u^3}{\partial x^i}\right)^2, & \text{if } i = j; \\ \left(\frac{\partial u^j}{\partial x^i} + \frac{\partial u^i}{\partial x^j}\right) - \frac{\partial u^1}{\partial x^i}\frac{\partial u^1}{\partial x^j} - \frac{\partial u^2}{\partial x^i}\frac{\partial u^2}{\partial x^j} - \frac{\partial u^3}{\partial x^i}\frac{\partial u^3}{\partial x^j}, & \text{if } i \neq j; \end{cases}$$

and

$$\mathcal{J} = \frac{1}{1!}\delta^i_j \varepsilon^j_i; \tag{44}$$

$$\mathcal{J}\mathcal{J} = \frac{1}{2!}\delta^{ij}_{kl}\varepsilon^l_j \varepsilon^k_i; \tag{45}$$

$$\mathcal{J}\mathcal{J}\mathcal{J} = \frac{1}{3!}\delta^{ijk}_{rst}\varepsilon^t_k \varepsilon^s_j \varepsilon^r_i, \tag{46}$$

where δ^{ij}_{kl} and δ^{ijk}_{rst} are the generalized Kronecker delta [46, 47]. Thus, the relationship between dV (differential of volume in the reference configuration) and dv (differential of volume in the actual configuration) can be described in the following way:

$$dV = (-8\mathcal{J}\mathcal{J}\mathcal{J} + 4\mathcal{J}\mathcal{J} - 2\mathcal{J} + 1)^{\frac{1}{2}}dv. \tag{47}$$

Taking into account that the yield region can be characterized as a transition region. This region, in fact, can be treated as a singularity region. As the continuity axiom holds a suitable asymptotic behavior [48–50], the concept of transition phenomenon has an asymptotic nature, and can be treated by limiting approaches. Physically, the transition region (transition/yielding) can be noted when a substance loses part of their original intrinsic properties, and from this point a new set of properties are raised in this substance, resulting in a new behavior. For example, in an elasto-plastic transition behavior, the transition state is related to a strain ellipsoid degeneracy, in a geometric sense, which can be transformed to an infinite cylinder, point or a pair of planes [51].

Considering the abovementioned lines, it can be imagined the nonlinear viscoelastic region as a direct mapping image from the linear viscoelastic region. Thus, the Jacobian ($\tilde{J}$), of this mapping, should translate the singularity on the transition region vicinity as an asymptotic extremely large deformation of the original microscopic element. In other words, taking into account Eq. (47), it can be seen that for a finite initial volume dV, with $\tilde{J} \to 0$ (singularity on the neighbor of the transition region) implicates in $dv \to \infty$, and in this way, it can write for the transition region the following condition:

$$8\mathcal{J}\mathcal{J}\mathcal{J} - 4\mathcal{J}\mathcal{J} + 2\mathcal{J} = 1, \tag{48}$$

or in an asymptotic sense on the transition region (t.r.)

$$\lim_{\underline{\underline{\varepsilon}} \to \text{t.r.}} (8\mathcal{J}\mathcal{J}\mathcal{J} - 4\mathcal{J}\mathcal{J} + 2\mathcal{J}) = 1. \tag{49}$$

It is important to stand out that this condition was stated in terms of the strain invariants, which can be rewritten in terms of the stress invariants or in terms of energy, by the constitutive relationships. Note the difficulty involved in expressing the strain tensor invariants in terms of the stress tensor invariants, due to the constitutive model's form.

Remark 1. *If it is taken the time rate of* Eq. (48), *one has*

$$4\dot{\overline{\mathcal{J}\mathcal{J}\mathcal{J}}} - 2\dot{\overline{\mathcal{J}\mathcal{J}}} + \dot{\mathcal{J}} = 0;$$

$$\therefore 4\mathcal{J}\mathcal{J}\mathcal{J}\underline{\underline{\varepsilon}}^{-T} : \dot{\underline{\underline{\varepsilon}}} - 2\left[\mathcal{J}\mathcal{J}tr\left(\dot{\underline{\underline{\varepsilon}}}\right) - \underline{\underline{\varepsilon}} : \dot{\underline{\underline{\varepsilon}}}\right] + tr\left(\dot{\underline{\underline{\varepsilon}}}\right) = 0;$$

or

$$4\mathcal{J}\mathcal{J}\mathcal{J}\underline{\underline{\varepsilon}}^{-T} : \underline{\underline{D}} - 2\left[\mathcal{J}\mathcal{J}tr\left(\underline{\underline{D}}\right) - \underline{\underline{\varepsilon}} : \underline{\underline{D}}\right] + tr\left(\underline{\underline{D}}\right) = 0. \tag{50}$$

Supposing the case that the strain rate is obtained as a response to known stress load, returning to Eq. (36), it follows:

$$\overset{\odot}{\underline{\underline{\Pi}}} + \theta_1^{-1}\underline{\underline{\Pi}} = \frac{G(\eta_\nu + \eta_\mu)}{\eta_\nu \eta_\mu}\dot{\underline{\underline{\Gamma}}} + \overset{\odot}{\dot{\underline{\underline{\Gamma}}}}, \tag{51}$$

or

$$\overset{\odot}{\dot{\underline{\underline{\Gamma}}}} + \phi' \dot{\underline{\underline{\Gamma}}} = \underline{\underline{\psi}}', \tag{52}$$

with

$$\underline{\underline{\psi}}' = \overset{\odot}{\underline{\underline{\Pi}}} + \theta_1^{-1}\underline{\underline{\Pi}};$$

$$\phi' = \frac{G(\eta_\nu + \eta_\mu)}{\eta_\mu \eta_\nu}.$$

Thus, one has

$$\vartheta' \overset{\odot}{\dot{\underline{\underline{\Gamma}}}} + \vartheta' \phi' \dot{\underline{\underline{\Gamma}}} = \overset{\odot}{\overline{\vartheta' \dot{\underline{\underline{\Gamma}}}}} = \vartheta' \underline{\underline{\psi}}' \therefore \overset{\odot}{\underline{\underline{\Theta}}} = \underline{\underline{\psi}}, \tag{53}$$

where

$$\vartheta'(t) = exp\left(\int \phi' dt\right);$$

$$\underline{\underline{\psi}}(t) = \vartheta' \underline{\underline{\psi}}';$$

$$\underline{\underline{\Theta}} = \vartheta' \dot{\underline{\underline{\Gamma}}}.$$

Note that

$$\dot{\overline{\underline{\underline{F}}\left(\underline{\underline{F}}^{-1}\underline{\underline{\Theta}}\underline{\underline{F}}^{-T}\right)\underline{\underline{F}}^T}} = \dot{\underline{\underline{F}}}\left(\underline{\underline{F}}^{-1}\underline{\underline{\Theta}}\underline{\underline{F}}^{-T}\right)\underline{\underline{F}}^T + \underline{\underline{F}}\dot{\overline{\left(\underline{\underline{F}}^{-1}\underline{\underline{\Theta}}\underline{\underline{F}}^{-T}\right)}}\underline{\underline{F}}^T + \underline{\underline{F}}\left(\underline{\underline{F}}^{-1}\underline{\underline{\Theta}}\underline{\underline{F}}^{-T}\right)\dot{\underline{\underline{F}}}^T;$$

$$\therefore \dot{\underline{\underline{\Theta}}} = \underline{\underline{L}}\underline{\underline{\Theta}} + \underline{\underline{F}}\dot{\overline{\left(\underline{\underline{F}}^{-1}\underline{\underline{\Theta}}\underline{\underline{F}}^{-T}\right)}}\underline{\underline{F}}^T + \underline{\underline{\Theta}}\underline{\underline{L}}^T;$$

$$\therefore \dot{\underline{\underline{\Theta}}} - \underline{\underline{L}}\underline{\underline{\Theta}} - \underline{\underline{\Theta}}\underline{\underline{L}}^T = \underline{\underline{F}}\dot{\overline{\left(\underline{\underline{F}}^{-1}\underline{\underline{\Theta}}\underline{\underline{F}}^{-T}\right)}}\underline{\underline{F}}^T, \tag{54}$$

and in this way, from Eq. (53), it follows:

$$\overset{\odot}{\underline{\underline{\Theta}}} = \underline{\underline{F}}\dot{\overline{\left(\underline{\underline{F}}^{-1}\underline{\underline{\Theta}}\underline{\underline{F}}^{-T}\right)}}\underline{\underline{F}}^T = \underline{\underline{\psi}};$$

$$\therefore \dot{\overline{\left(\underline{\underline{F}}^{-1}\underline{\underline{\Theta}}\underline{\underline{F}}^{-T}\right)}} = \underline{\underline{F}}^{-1}\underline{\underline{\psi}}\underline{\underline{F}}^{-T};$$

$$\therefore \underline{\underline{\Theta}}(t) = \underline{\underline{F}}(t)\left(\int_0^t \underline{\underline{F}}^{-1}(t')\underline{\underline{\psi}}(t')\underline{\underline{F}}^{-T}(t')\ dt'\right)\underline{\underline{F}}^T(t);$$

$$\therefore \dot{\underline{\underline{\Gamma}}}(t) = \frac{1}{\vartheta'(t)}\underline{\underline{F}}(t)\left(\int_0^t \underline{\underline{F}}^{-1}(t')\underline{\underline{\psi}}(t')\underline{\underline{F}}^{-T}(t')\ dt'\right)\underline{\underline{F}}^T(t),$$

or in other words

$$\underline{\underline{D}}(t) = \frac{G}{\eta_\mu \vartheta'(t)} \underline{\underline{F}}(t) \left(\int_0^t \underline{\underline{F}}^{-1}(t') \underline{\underline{\psi}}(t') \underline{\underline{F}}^{-T}(t')\ dt' \right) \underline{\underline{F}}^T(t);$$

$$= \frac{G}{\eta_\mu \vartheta'(t)} \underline{\underline{F}}(t) \left[\int_0^t \underline{\underline{F}}^{-1}(t') \vartheta'(t') \left(\overset{\odot}{\underline{\underline{\Pi}}}(t') + \theta_1^{-1} \underline{\underline{\Pi}}(t') \right) \underline{\underline{F}}^{-T}(t')\ dt' \right] \underline{\underline{F}}^T(t);$$

$$\therefore \underline{\underline{D}}(t) = \frac{G}{\eta_\mu \vartheta'(t)} \underline{\underline{F}}(t) \left\{ \int_0^t \underline{\underline{F}}^{-1}(t') \exp\left(\int \frac{G(\eta_\nu + \eta_\mu)}{\eta_\mu \eta_\nu} dt' \right) \right.$$

$$\left[\overline{\left(\frac{1}{G} \right)} \underline{\underline{\tau}}(t') + G^{-1} \left(\underline{\underline{\tau}}_{,t'}(t') + \underline{\underline{\tau}}(t') \cdot \underline{\mathfrak{v}}(t') - \underline{\underline{L}}(t') \underline{\underline{\tau}}(t') - \underline{\underline{\tau}}(t') \underline{\underline{L}}^T(t') \right) \right.$$

$$\left. \left. + \frac{\underline{\underline{\tau}}(t')}{\eta_\nu} \right] \underline{\underline{F}}^{-T}(t')\ dt' \right\} \underline{\underline{F}}^T(t), \tag{55}$$

where $\underline{\underline{L}} = \underline{\mathfrak{v}}$ ("Eulerian gradient"), and where $\underline{\mathfrak{v}}$ is the velocity field. Consequently,

$$\underline{\underline{\varepsilon}}(t) = \frac{G}{\eta_\mu} \int_0^t \frac{1}{\vartheta'(t')} \underline{\underline{F}}(t') \left\{ \int_0^{t'} \underline{\underline{F}}^{-1}(t'') \exp\left(\frac{G(\eta_\nu + \eta_\mu)}{\eta_\mu \eta_\nu} t'' \right) \right.$$

$$\left[G^{-1} \left(\underline{\underline{\tau}}_{,t''}(t'') + \underline{\underline{\tau}}(t'') \cdot \underline{\mathfrak{v}}(t'') - \underline{\underline{L}}(t'') \underline{\underline{\tau}}(t'') - \underline{\underline{\tau}}(t'') \underline{\underline{L}}^T(t'') \right) \right.$$

$$\left. \left. + \eta_\nu^{-1} \underline{\underline{\tau}}(t'') \right] \underline{\underline{F}}^{-T}(t'')\ dt'' \right\} \underline{\underline{F}}^T(t') dt', \tag{56}$$

since λ_μ and λ_ν are taken as constants (pre-transition/yield region). In this way, the criteria Eq. (48) and/or Eq. (50) can be appropriately analyzed for a stress load excitation.

5. Numerical results

In this section, some illustrative numerical results are presented. Two types of rheological tests are analyzed, the constant shear rate test (CR) and the constant shear stress test (CS). In both cases it is necessary to define consistent initial conditions in relation to physical and mathematical principles. The finite difference method was implemented in MATLAB, with the following set of parameters: $\aleph = 1$, $\eta_0 = 0.08\ \text{Pa.s}$, $\alpha_1 = 5$, $\alpha_2 = \alpha_3 = 0.5$, $k_\nu \varsigma_\nu = 10^3\ \text{J/m}^3$, $k_\mu \varsigma_\mu = 10^2\ \text{J/m}^3$, $\beta_\nu = 5$, $\beta_\mu = 3$, $m = 10^{-6}$, $K_\nu^* = 0.001\ \text{Kg m}^{-1}\text{K}^{-1}$, $K_\mu^* = 0.01\ \text{Kg m}^{-1}\text{K}^{-1}$, $G_0 = 6.5\ \text{Pa}$, $t_\nu \varsigma_\nu = 10^4\ \text{s}^{-1}$ and $t_\mu \varsigma_\mu = 10^3\ \text{s}^{-1}$. It is important to point out that for both tests it was used $\Delta t = 10^{-3}$ s. It is also important to comment that time-steps $\Delta t = 10^{-4}$ s, 10^{-5} s and 10^{-6} s were analyzed, but modifications in responses were not noted. The tolerance used for the Newton's method was 10^{-6}. Numerical results for real thixotropic substances (numerical/experimental comparative responses, regularization methods associated to nonlinear identification parameter problem, etc.) can be found in [52].

5.1. Constant shear rate test

In this section, it is considered the constant shear rate test, that is taken into account load conditions as $\dot{\gamma}(t) = H(t)\dot{\gamma}_0$, where $H(t)$ represents the standard Heaviside function and $\dot{\gamma}_0$ it is a positive real constant. It is reasonable to think that in the begin of the test, the micro-structure is in a fully structured state ($\lambda_\mu(0) = \lambda_\nu(0) = 1$).

Figure 3 shows an interesting aspect, for the loads $\dot{\gamma}_0 = 10^{-2}\ \mathrm{s}^{-1}$ and $10^{-1}\ \mathrm{s}^{-1}$ the behavior is close to a purely viscoelastic response. These behaviors are in agreement with **Figures 4** and **5**, for the same loads. Note that these two shear rates correspond to a low level of modification in λ_μ and λ_ν. In these both cases, the stress of steady state is the maximum stress reached.

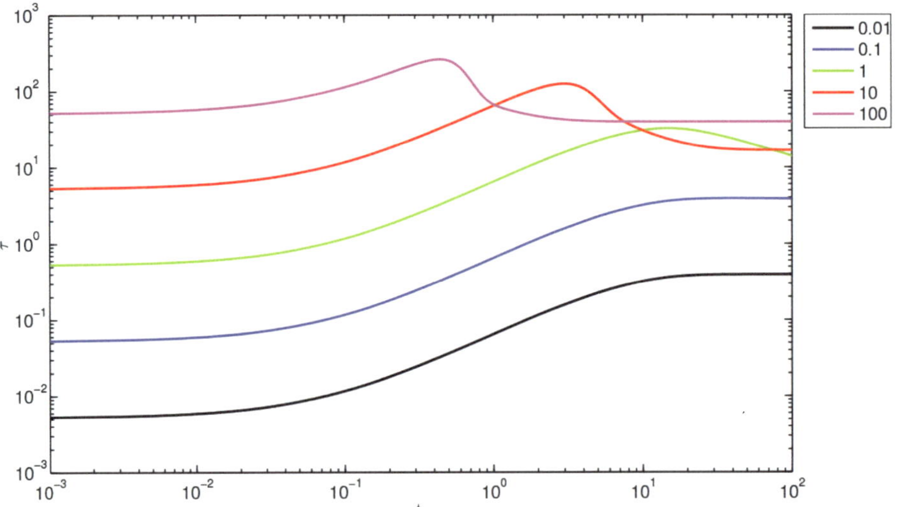

Figure 3.
τ vs. t.

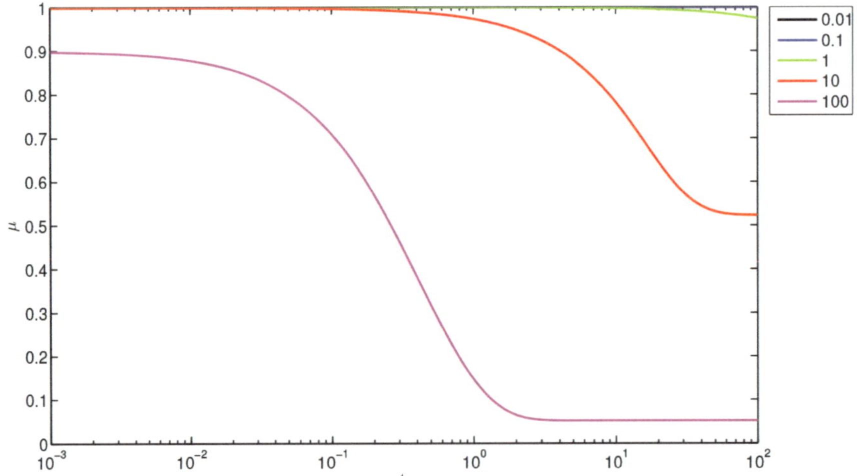

Figure 4.
λ_μ vs. t.

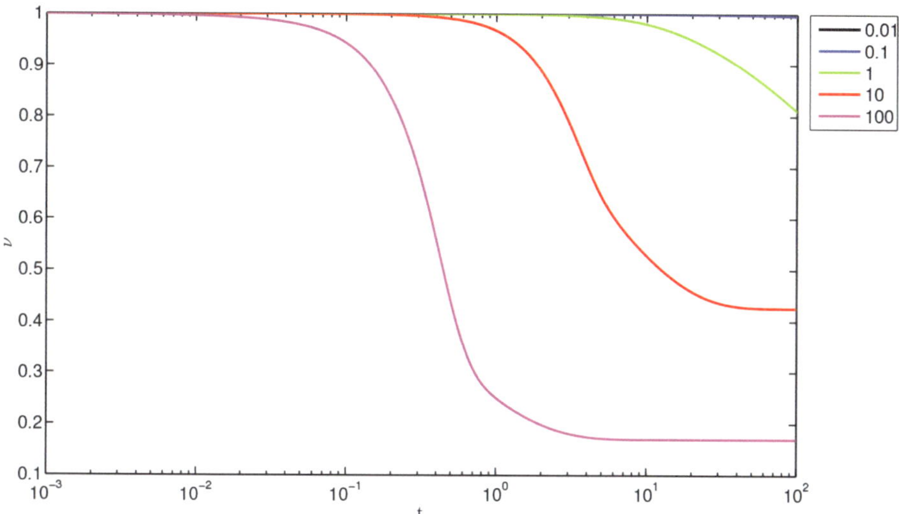

Figure 5.
λ_ν vs. t.

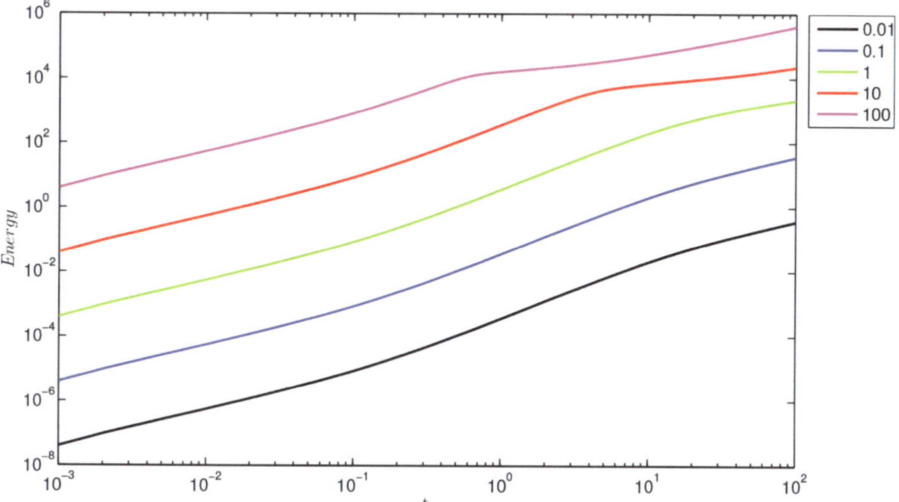

Figure 6.
Energy vs. t.

It is important to point out that for $\dot{\gamma}_0 = 1\ \text{s}^{-1}$ the thixotropic effect can ever be noted in stress response (**Figure 3**). The loads $\dot{\gamma}_0 = 10\ \text{s}^{-1}$ and 100 s^{-1} show a typical thixotropic behavior (**Figure 3**). These responses are related to considerable modifications in the structural nature of the substance, as can be seen in **Figures 4** and 5. In these both cases, the maximum stress occurs before the steady state. It can be proved, in a theoretical/mathematical sense, that the point where the maximum breakdown rate occurs is before the stress peak. This fact is in agreement with expected behavior of thixotropic substance.

Other interesting result is related to the energy behavior of the thixotropic substance (**Figures 6** and 7). Note the presence of a specific slope change in the region where higher rates of decrease on the micro-structures (λ_μ, λ_μ) levels occurred. The same behavior can be observed on experimental results. In fact, this can be explained, in a theoretical sense, via the following relationship

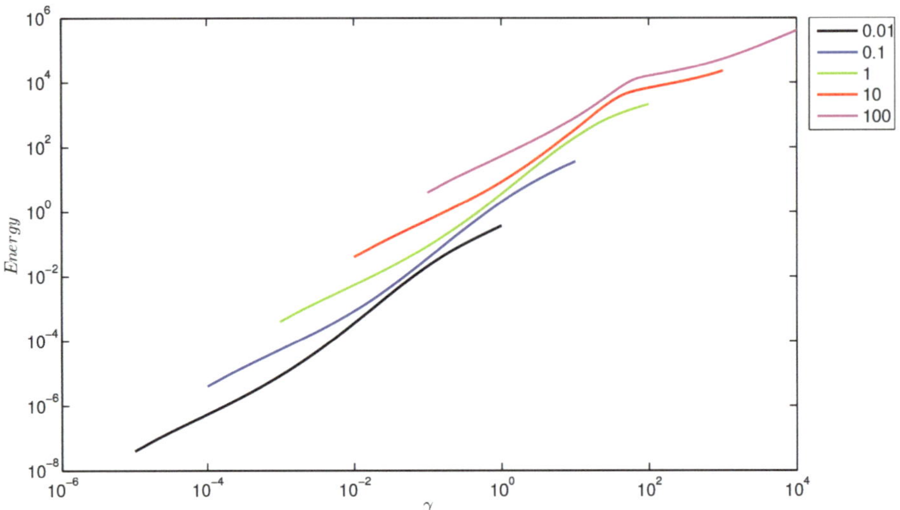

Figure 7.
Energy vs. γ.

$$Power = k_\mu \varsigma_\mu \frac{(1-\lambda_\mu)^\beta}{\lambda_\mu} - t_\mu \varsigma_\mu \frac{\dot{\lambda}_\mu}{\lambda_\mu} + k_\nu \varsigma_\nu \frac{(1-\lambda_\nu)^\beta}{\lambda_\nu} - t_\nu \varsigma_\nu \frac{\dot{\lambda}_\nu}{\lambda_\nu}, \tag{57}$$

that comes from the rate Eqs. (11) and (12).

5.2. Constant stress test

This chapter presents some points and results for the constant shear stress test. In this sense, it is considered loads as $\tau(t) = H(t)\tau_0$, where $H(t)$ represents the standard Heaviside function and τ_0 is a positive real constant. It is assumed that in the begin of the test, the micro-structure is in a fully structured state ($\lambda_\mu(0) = \lambda_\nu(0) = 1$).

It can be seen in **Figure 8**, for the loads $\tau_0 = 10, 20, 30, 40, 50$ and 100 Pa, a typical behavior for thixotropic substances under low level of constant stress loads.

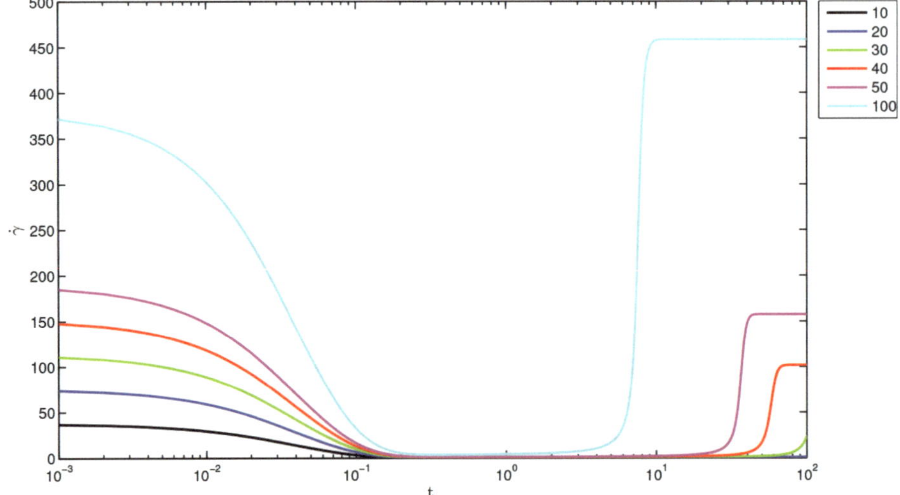

Figure 8.
$\dot{\gamma}$ vs. t.

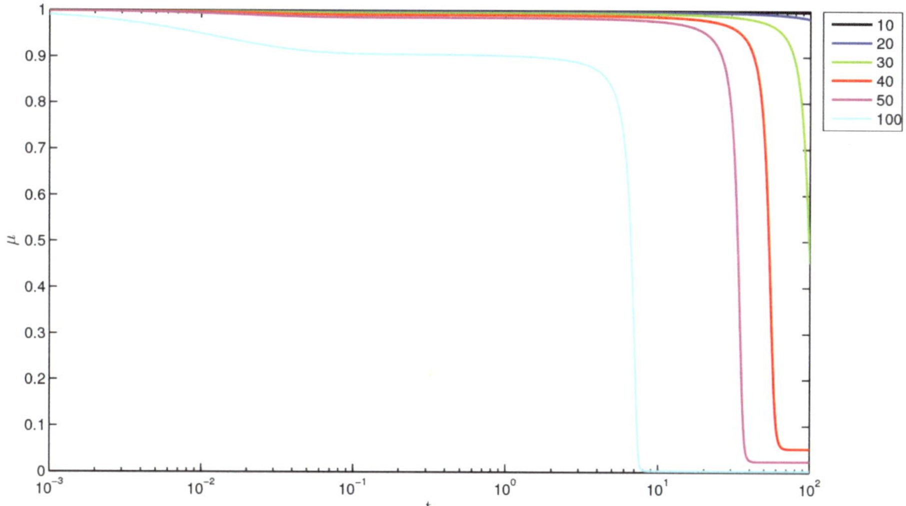

Figure 9.
λ_μ vs. t.

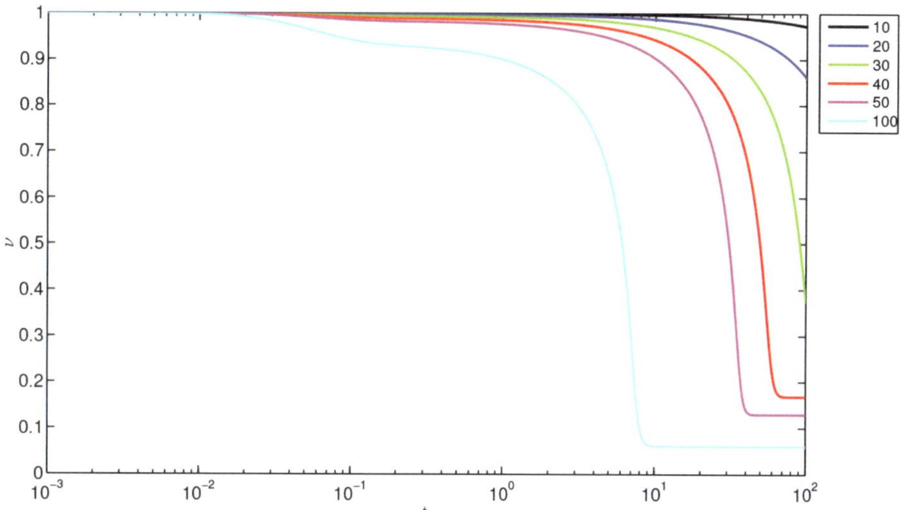

Figure 10.
λ_ν vs. t.

In this test, it can be seen the "Avalanche effect," and their relation with the micro-structural level (λ_μ, λ_ν) of the substance (**Figures 9** and **10**). In these cases, it can be observed the strain rate decrease, related to an increase of the construction parcel (cp_μ) and decrease of the destruction parcel (dp_μ) of λ_μ, as can be seen in **Figures 11** and **12** respectively, where

$$cp_\mu = \frac{1}{t_\mu}\left(k_\mu(1-\lambda_\mu)^{\beta_\mu}\right); \tag{58}$$

$$dp_\mu = \frac{\left(K_\mu^* \lambda_\mu^6 \theta \ddot{\gamma} + \tau_\mu\right)\lambda_\mu \dot{\gamma}}{\varsigma_\mu}. \tag{59}$$

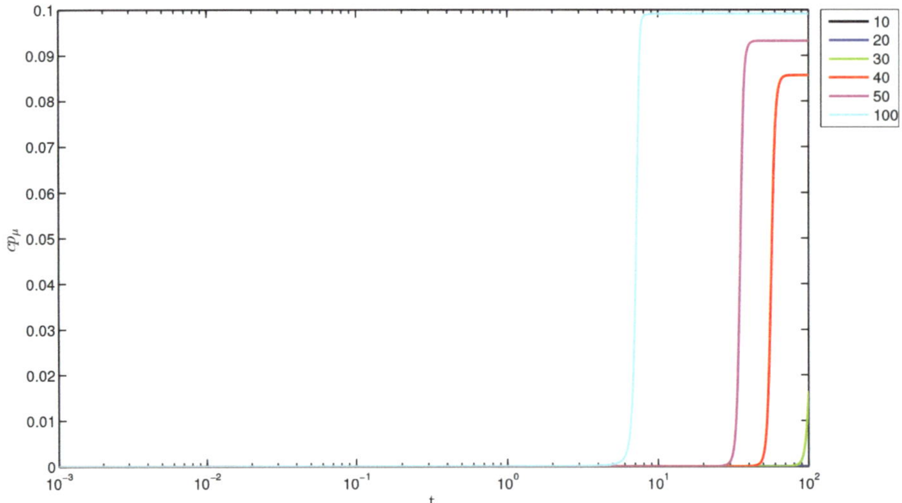

Figure 11.
cp_μ vs. t.

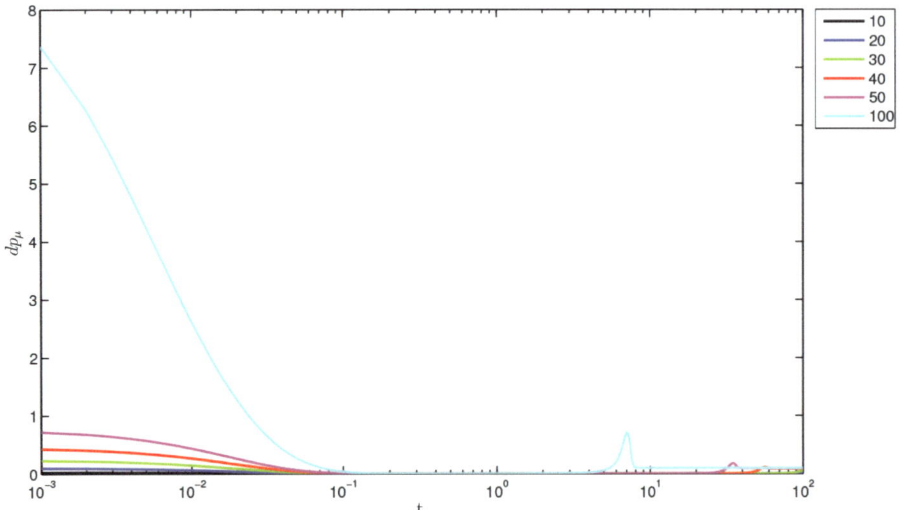

Figure 12.
dp_μ vs. t.

For the Maxwell element $(._\nu)$, in the same region (the strain rate decreases), cp_ν is close to null value and dp_ν is increasing, but this increase level is not sufficient to stop the decreasing process of $\dot{\lambda}$. A posterior increase of $\dot{\gamma}$ is due to an abrupt decrease of linkages number $(\lambda_\mu, \lambda_\nu)$ before the steady state.

It is important to note the relationship between the structural nature and the behavior of the substance (**Figures 9** and **10**). Small changes in the values of λ_μ and λ_ν were detected for stress loads 10 and 20 Pa, in relation to the others. In these two cases $\dot{\gamma}$ presents an nonincreasing behavior.

Other interesting result is presented in **Figure 13**, where it can be seen the energy behavior along the test. Note a changing on slope of the energy lines, tending to horizontal slope, in the $\dot{\gamma}$ decreasing region and another one related to the $\dot{\gamma}$ increasing region.

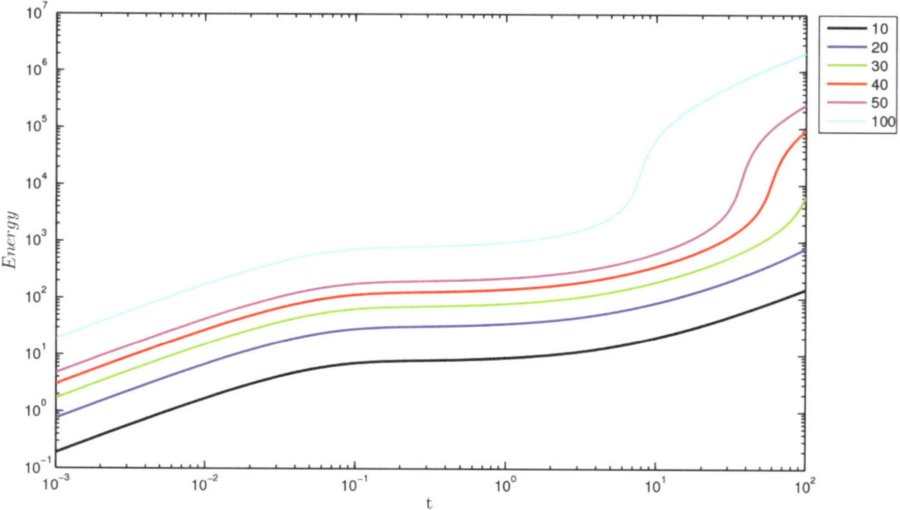

Figure 13.
Energy vs. *t*.

6. Concluding remarks

The main objective of this chapter is to investigate the constitutive model for thixotropic fluids based on [8] related to the existence of two different types of micro-structure, and their consistence in some rheological tests. The model and analysis presented here are based on well-established physics principles. In this sense, it is important to stand out some nonstandard points of the approach presented in this work for thixotropic modeling in respect to the others models. In this sense, it is important to stand out some points:

- the shear modulus ($G(\lambda_\nu)$), the viscosity coefficients ($\eta_\mu(\lambda_\mu)$ and $\eta_\nu(\lambda_\nu)$) and their dependences from the two different micro-structure types are considered in the development of the constitutive model (Eqs. (9)–(14));
- the set of the rate equations (Eqs. (11)–(12)) are related to well-established physical principles as the "reptation" model and the "Smoluchowski" theory of coagulation. It is important to point out that in the "Smoluchowski" theory of coagulation the effect of the Brownian motion is clearly taken into account;
- the thermodynamic consistence of the model was analyzed (Section 3);
- a theoretical criterium for the transition region based on the strain gradient mapping degeneration was discussed and exploited (Section 4), however it is important to stand out the necessity of more discussions and analysis on the relationships (Eqs. (48) and (56)) taking into account some additional characteristics (as temperature, ...) and their consequences. These characteristics shall be considered in future work goals;
- the illustrative numerical examples attest the capability of the model to predict the expected behavior of real thixotropic substances under some typical rheological tests (Section 5).

It is also important to comment that the developed ideas and presented in this work can be easily extended to approaches including more than two types of microstructure (related to each specific relaxation mechanism) that are taken into account in the constitutive equation system. It is clear that the complexity level of the considered substance determines the necessity of these incorporations. In this context, it can be noted the versatility of the approach exposed here, presenting some interesting perspectives on thixotropic modeling that can be more explored in future works.

Symbology

τ	shear stress
γ	shear strain
G	shear modulus
η	dynamic viscosity
λ	structural parameter
$\mathfrak{G}\dot{\gamma}$	functional associated to structural parameter time evolution
$\delta^{i\ldots}_{r\ldots}$	the generalized Kronecker delta
$\dot{}$	total time derivative
$\ddot{}$	double total time derivative
$\odot$	Oldroyd time derivative
ν	associated to the Maxwellian mechanism
μ	associated to the pure viscous mechanism
$\bar{}$	first-order tensor
$=$	second-order tensor
$\cdot$	inner product
$:$	double inner product

Author details

Hilbeth P. Azikri de Deus[1*] and Mikhail Itskov[2]

1 Nucleus of Applied and Theoretical Mechanics – NuMAT, UTFPR, Brazil

2 Department of Continuum Mechanics, RWTH – Aachen University, Germany

*Address all correspondence to: azikri@utfpr.edu.br

References

[1] Zhang X, Li W, Gong X. Thixotropy of MR shear-thickening fluids. Smart Materials and Structures. 2010; **19**(125012):1-6

[2] El-Gendy H, Alcoutlabi M, Jemmett M, Deo M, Magda J, Venkatesan R, Montesi A. The propagation of pressure in a gelled waxy oil pipeline as studied by particle imaging velocimetry. AICHE Journal. 2012;**58**(3):302-311

[3] Nguyen Q, Boger D. Thixotropic behaviour of concentrated bauxite residue suspensions. Rheologica Acta. 1985;**24**:427-437

[4] Mujumdar A, Beris AN, Metzner AB. Transient phenomena in thixotropic systems. Journal of Non-Newtonian Fluid Mechanics. 2002;**102**:157-178

[5] de Souza Mendes PR. Modeling the thixotropic behavior of structured fluids. Journal of Non-Newtonian Fluid Mechanics. 2009;**164**:66-75

[6] Swift DL, Friedlander SK. The coagulation of hydrosols by Brownian motion and laminar shear flow. Journal of Colloid Science. 1964;**19**: 621-647

[7] Yaghouti MR, Rezakhanlou F, Hammond A. Coagulation, diffusion and the continuous Smoluchowski equation. Stochastic Processes and their Applications. 2009;**119**:3042-3080

[8] Azikri de Deus HP, Negrão CRO, Franco AT. The modified Jeffreys model approach for elasto-viscoplastic thixotropic substances. Physics Letter A. 2016;**380**:585-595

[9] Hammond A, Rezakhanlou F. Kinetic limit for a system of coagulating planar Brownian particles. Journal of Statistical Physics. 2006;**124**:997-1040

[10] Niethammer B, Velázquez JJL. Self-similar solutions with fat tails for Smoluchowski's coagulation equation with locally bounded kernels. Communications in Mathematical Physics. 2013;**318**:502-532

[11] Pokrovskii VN. A justification of the reptation-tube dynamics of a linear macromolecule in the mesoscopic approach. Physica A. 2006;**366**:88-106

[12] Öttinger HC. A thermodynamically admissible reptation model for fast flows of entangled polymers. Journal of Rheology. 1999;**43**:1461-1493

[13] Mewis J. Thixotropy: A general review. Journal of Non-Newtonian Fluid Mechanics. 1979;**6**:1-20

[14] Barnes H. Thixotropy: A review. Journal of Non-Newtonian Fluid Mechanics. 1997;**70**:1-33

[15] Toorman EA. Modeling the thixotropic behaviour of dense cohesive sediment suspensions. Rheologica Acta. 1997;**36**:56-65

[16] Ritter RA, Batycky JP. Numerical prediction of the pipeline flow characteristics of thixotropic liquids. SPE Journal. 1967;7:369-376

[17] Dullaert K, Mewis J. A structural kinetics model for thixotropy. Journal of Non-Newtonian Fluid Mechanics. 2006; **139**:21-30

[18] de Souza Mendes PR. Thixotropic elasto-viscoplastic model for structured fluids. Soft Matter. 2011;7:2471-2483

[19] Azikri de Deus HP, Dupim GSP. Over strucutural nature of the thixotropic fluids behavior. Applied Mathematical Sciences. 2012;**6**: 6871-6889

[20] Azikri de Deus HP, Dupim GSP. On behavior of the thixotropic fluids. Physics Letters A. 2013;**337**:478-485

[21] Azikri de Deus HP, Dupim GSP. Some aspects over thixotropic fluid behavior. Advanced Materials Research. 2013;**629**:623-634

[22] Ardakani HA, Mitsoulis E, Hatzikiriakos SG. Thixotropic flow of toothpaste through extrusion dies. Journal of Non-Newtonian Fluid Mechanics. 2011;**166**:1262-1271

[23] Oliveira GM, Rocha LLV, Franco AT, Negrão COR. Numerical simulation of the start-up of Bingham fluid flows in pipelines. Journal of Non-Newtonian Fluid Mechanics. 2010;**165**:1114-1128

[24] Mewis J, Wagner NJ. Thixotropy. Advances in Colloid and Interface Science. 2009;**147–148**:214-227

[25] Dullaert K, Mewis J. Thixotropy: Build-up and breakdown curves during flow. Journal of Rheology. 2005;**49**(6): 1213-1230

[26] Potanin A. 3D simulations of the flow of thixotropic fluids, in large-gap Couette and vane-cup geometries. Journal of Non-Newtonian Fluid Mechanics. 2010;**165**:299-312

[27] Derksen JJ, Prashant NV. Simulations of complex flow of thixotropic liquids. Journal of Non-Newtonian Fluid Mechanics. 2009;**160**: 65-75

[28] Barnes HA. The yield stress – A review or 'panta rei' – Everything flows? Journal of Non-Newtonian Fluid Mechanics. 1999;**81**:133-178

[29] Patil PD, Feng JJ, Hatzikiriakos SG. Constitutive modeling and flow simulation of polytetrafluoroethylene (PTFE) paste extrusion. Journal of Non-Newtonian Fluid Mechanics. 2006;**139**: 44-53

[30] Marrucci G. The free energy constitutive equation for polymer solutions from the dumbell model. Transactions. Society of Rheology. 1972; **16**:321-330

[31] Marrucci G, Titomanlio G, Sarti GC. Testing of a constitutive equation for entangled networks by elongational and shear data of polymer melts. Rheologica Acta. 1973;**12**:269-275

[32] Silva TABP. Aálise do Módulo de Cisalhamento Associado a Modelo de Jeffreys Modificado (in Portuguese). Master Dissertation. UTFPR, PPGEM, Brazil; 2017

[33] Rajagopal KR, Srinivasa AR. On thermomechanical restrictions of continua. Proceedings of the Royal Society of London A. 2004;**406**:631-651

[34] Rajagopal KR, Srinivasa AR. Inelastic behavior of materials. I. Theoretical underpinnings. International Journal of Plasticity. 1998; **14**:945-967

[35] Rajagopal KR, Srinivasa AR. Inelastic behavior of materials II. Inelastic response. International Journal of Plasticity. 1998;**14**:967-995

[36] Rajagopal KR, Srinivasa AR. A thermodynamic framework for rate-type fluid models. Journal of Non-Newtonian Fluid Mechanics. 2000;**88**: 207-227

[37] Marze S, Guillermic RM, Saint-James A. Oscillatory rheology of aqueous foams: Surfactant, liquid fraction, experimental protocol and aging effects. Soft Matter. 2009;**5**(9): 1937

[38] Mohan L, Pellet C, Cloitre M, Bonnecaze R. Local mobility and microstructure in periodically sheared soft particle glasses and their connection to macroscopic rheology. Journal of Rheology. 2013;**57**(3):1023

[39] Mason TG, Bibette J, Weitz DA. Yielding and flow of monodisperse emulsions. Journal of Colloid and Interface Science. 1996;**179**(179): 439-448

[40] Souza Mendes PR, Thompson RL, Alicke AA, Leite RT. The quasilinear large-amplitude viscoelastic regime and its significance in the rheological characterization of soft matter. Journal of Rheology. 2014;**58**(2):537-561

[41] Walls HJ, Caines SB, Sanchez AM, Khan SA. Yield stress and wall slip phenomena in colloidal silica gels. Journal of Rheology. 2003;**47**(4):847

[42] Andrade DEV, Takii BA, Franco AT, Negrão CRO. The influence of the initial cooling condition on the flow curve of waxy crude oil. In: 15th Brazilian Congress of Thermal Sciences and Engineering. Belém: ABCM; 2014

[43] Divoux T, Barentin C, Manneville S. Stress overshoot in a simple yield stress fluid: An extensive study combining rheology and velocimetry. Soft Matter. 2011;7:9335-9349

[44] Fernandes RR, Andrade DEV, Franco AT, Negrão COR. Sampling methodology for rheological tests of drilling fluids: A study of the aging time and preshearing. In: 15th Brazilian Congress of Thermal Sciences and Engineering. Belém; 2014

[45] Knauss WG, Zhu W. Nonlinearly viscoelastic behavior of polycarbonate. I response under pure shear. Mechanics of Time-Dependent Materials. 2002;**6**: 231-269

[46] Agarwa DC. Tensor Calculus and Riemannian Geometry. Meerut, India: Krishna Prakashan Media Ltd.; 2007

[47] Lovelock D, Rund H. Tensors, Differential Forms, and Variational Principles. New York, USA: Dover Publications, INC.; 1989

[48] Seth BR. Elastic-plastic transition in shells and tubes under pressure. Journal of Applied Mathematics and Mechanics. 1963;**43**:345-351

[49] Seth BR. Generalized strain and transition concepts for elasti-plastic deformation, creep and relaxation. In: Proceedings of the Eleventh International Congress of Applied Mechanics; 1966. pp. 383–389

[50] Seth BR. Transition Concept in Continuum Mechanics. Corvallis, USA: Seminar Lectures delivered in Oregon State University (unpublished); 1968

[51] Purushothama CM. Elastic-plastic transition. Journal of Applied Mathematics and Mechanics. **45**: 401-408

[52] Morinigo V Jr. Verificação Teórico-numéica de Modelo Constitutivo Aplicado a Fluidos Tixotrópicos Compostos por Duas Estruturas Distintas (in Portuguese). TCC, Federal University of Technology-Paraná – UTFPR, Brazil; 2017

Chapter 4

Practical Methods for Online Calculation of Thermoelastic Stresses in Steam Turbine Components

Mariusz Banaszkiewicz and Janusz Badur

Abstract

This chapter presents two practical methods of thermoelastic stress calculation suitable for application in online monitoring systems of steam turbines. Both methods are based on the Green function and Duhamel integral and consider the effect of variable heat transfer coefficient and material physical properties on thermal stresses. This effect is taken into account either by using an equivalent steam temperature determined with a constant heat transfer coefficient or by applying an equivalent Green's function determined with variable heat transfer coefficient and physical properties. The effectiveness of both methods was shown by comparing their predictions with the results of exact three-dimensional (3D) calculations of a steam turbine valve.

Keywords: thermoelastic stress, steam turbine, online calculation

1. Introduction

The increasing demand for a higher thermal efficiency and a higher operational flexibility of modern power generation plants results in severe mechanical and thermal loading. Modern energy markets put a requirement of a high operational flexibility of power plant units [1]. The flexible operation generates elevated loads and stresses in turbine components leading to material damage due to thermo-mechanical fatigue. In order to control the fatigue damage, thermal stresses should be computed online and limited to permissible values [2].

In power generation industry, the approach based on the Duhamel integral and the Green functions has been widely used for thermal stress calculations [3–6]. Such an approach has also been adopted for monitoring power boiler operation [7] and can also be used for calculating stress intensity factors at transient thermal loads [8].

The major issue in using Green's function and Duhamel's integral method in modeling transient thermal stresses in steam turbine components is the time dependence of material physical properties and heat transfer coefficients affecting proper evaluation of Green's functions and making the problem nonlinear. There are known approaches assuming the determination of the influence functions at constant values of these quantities [9]. The inclusion of temperature-dependent physical properties proposed by Koo et al. [10] relies on determining the weight functions for steady-state and transient operating conditions. A more important variation of the heat transfer coefficient can be taken into account by calculating the surface temperature using a reduced heat transfer model and employing Green's function to calculate a stress response to the step change of metal surface temperature [11]. However, numerical solution of a multi-dimensional heat transfer model for complex geometries present in steam turbines is complicated and time-consuming, and due to this, it cannot be used in online calculations. A full inclusion of time variability of the physical properties and heat transfer coefficients proposed by Zhang et al. [12] relies on the solution of nonlinear heat conduction problem by using artificial parameter method and superposition rule and replacing the time-dependent heat transfer coefficient with a constant value together with a modified fluid temperature. The effectiveness of the method has been proved by an example of a three-dimensional model of a pressure vessel of nuclear reactor.

This chapter presents and compares two alternative methods for steam turbine components both employing the Green functions and Duhamel's integral and considering the variations of material physical properties and heat transfer coefficient. The first method uses a numerical solution of one-dimensional (1D) heat conduction problem and determination of equivalent steam temperature for use with the Green function determined with a constant heat transfer coefficient. The second method employs the equivalent Green function determined with variable physical properties and heat transfer coefficient. The methods have been validated and compared for a steam turbine valve with the three-dimensional (3D) numerical solutions.

2. Thermoelasticity problem

The problem under consideration is heating up and cooling down of a steam turbine taking place during start-up and the subsequent shutdown. Heating and cooling of steam turbine components is caused by a hot steam flow through the turbine with varying temperature, pressure and velocity. Non-stationary heat transfer takes place via forced convection, and the thermal load is a primary load of the turbine. The rotor rotates with a constant rotational speed ω in a steady-state operation. Steam pressure and rotational body forces due to rotation are also considered in the model. Non-uniform temperature distribution in the component induces thermal stresses due to thermal expansion.

The material is assumed isotropic, and its physical and mechanical properties are temperature-dependent. The heating and cooling process is described by the linear theory of heat conduction and nonlinear convective boundary conditions. The resulting thermal stresses are determined from the solution of system of equations of uncoupled thermoelasticity. Due to the geometrical and loading symmetry of the turbine components like rotors or valve casings, the computational region is assumed axisymmetric. Mechanical loads are modeled as surface pressure and volumetric body force.

Based on the above assumptions, a boundary problem of heat conduction and thermoelasticity is formulated [13, 14] and solved by means of a finite element method (FEM) [15].

Knowing the axisymmetric temperature field in the component, a solution of the problem of stress state induced by it can be obtained. The equilibrium conditions written for stresses are transformed taking into account the relations between stresses, strains and displacements, to obtain differential equations with unknown displacements u and w in the radial r- and axial z-direction, respectively [16],

$$\nabla^2 u - \frac{u}{r^2} + \frac{1}{1-2\nu}\frac{\partial e}{\partial r} + \frac{\rho\omega^2 r}{G} = \frac{2(1+\nu)}{1-2\nu}\frac{\partial(\beta(T-T_0))}{\partial r} \tag{1}$$

$$\nabla^2 w + \frac{1}{1-2\nu}\frac{\partial e}{\partial z} = \frac{2(1+\nu)}{1-2\nu}\frac{\partial(\beta(T-T_0))}{\partial z}$$

where T is the temperature, T_0 is the initial temperature, ν is Poisson's ratio, β is the thermal expansion coefficient, ρ is the density and G is the shear modulus. In Eq. (1), the following definitions were used:

- Laplacian $\nabla^2 = \frac{\partial^2}{\partial r^2} + \frac{1}{r}\frac{\partial}{\partial r} + \frac{\partial^2}{\partial z^2}$
- Volumetric strain $e = \frac{\partial u}{\partial r} + \frac{u}{r} + \frac{\partial w}{\partial z}$

The system of partial differential equations (1) is solved with the following boundary conditions:

- At the outer cylindrical surfaces of the rotor

$$\sigma_r = p_b(z) \text{ is the pressure due to centrifugal forces of blades} \tag{2}$$

- At the outer surfaces of the rotor in contact with steam

$$\sigma_n = -p(z) \text{ is the steam pressure acting normal to the surface} \tag{3}$$

- At the inner surfaces of the casing in contact with steam

$$\sigma_n = -p(z) \text{ is the steam pressure acting normal to the surface} \tag{4}$$

- At the end face of the rotor or casing

$$w = 0 \text{ at } z = 0 \tag{5}$$

The relations between strains and displacements for an axisymmetric body in the cylindrical coordinate system are expressed as [16]

$$\begin{aligned} \varepsilon_r &= \frac{\partial u}{\partial r} \\ \varepsilon_\varphi &= \frac{u}{r} \\ \varepsilon_z &= \frac{\partial w}{\partial z} \\ \gamma_{rz} &= \frac{\partial u}{\partial z} + \frac{\partial w}{\partial r} \end{aligned} \tag{6}$$

Stress components are obtained from the solution of the constitutive equation of linear thermoelasticity [17], and for an axisymmetric body in the cylindrical coordinate system, they are given as [16]

$$\sigma_r = 2G\left\{\varepsilon_r + \frac{1}{1-2\nu}[\nu e - (1+v)\beta(T-T_0)]\right\}$$
$$\sigma_\varphi = 2G\left\{\varepsilon_\varphi + \frac{1}{1-2\nu}[\nu e - (1+v)\beta(T-T_0)]\right\} \tag{7}$$
$$\sigma_z = 2G\left\{\varepsilon_z + \frac{1}{1-2\nu}[\nu e - (1+v)\beta(T-T_0)]\right\}$$
$$\sigma_{rz} = G\gamma_{rz}$$

where
$G = \frac{E}{2(1+\nu)}$ is shear modulus and E is the Young modulus.

Eq. (7) is known as the Duhamel-Neumann relations and has a fundamental importance in thermal stress analysis within the validity of Hooke's law [18, 19].

3. Green's function method

It is assumed that the thermal stress model used in online calculations is based on Green's function and Duhamel's integral method. This method allows for fast calculation of thermal stresses at supervised areas for any changes of fluid temperature causing heating up or cooling down of an element. In steam turbine components, a reliable and accurate measurement of the surface temperature at the critical region is usually not possible; thus, for online monitoring purposes, the steam temperature measurement is used as a leading signal. Consequently, in heat transfer model, Fourier's boundary condition is employed [20]:

$$\lambda(r)\frac{\partial T(r,t)}{\partial n} = -\alpha(r)(T_{st}(t) - T(r,t)), \quad r = r_{out} \tag{8}$$

where $\lambda(r)$ is the metal thermal conductivity, $T(r, t)$ is the surface temperature, $T_{st}(t)$ denotes steam temperature, n is the normal direction and $\alpha(r)$ is the heat transfer coefficient.

Assuming that the steam temperature is described by Heaviside's function $H(t)$, the boundary condition (8) can be written in the form [10]:

$$\lambda(r)\frac{\partial X(r,t)}{\partial n} = -\alpha(r)(H(t) - X(r,t)) \tag{9}$$

where $X(r, t)$ is Green's function for temperature.

Using Green's function $X(r, t)$, we can calculate the metal temperature for arbitrary variation of steam temperature employing Duhamel's theorem [20]:

$$T(r,t) = \int_0^t X(r,t-\tau)\frac{\partial T_{st}(\tau)}{\partial \tau}d\tau \tag{10}$$

Considering elastic stresses, it can be assumed, according to the thermoelasticity theory, that stress distribution in an elastic body is a unique function of temperature distribution, and in this connection, the thermal stresses can be calculated using Duhamel's integral as

$$\sigma_{ij}(r,t) = \int_0^t G_{ij}(r,t-\tau)\frac{\partial T_{st}(\tau)}{\partial \tau}d\tau \tag{11}$$

where $G_{ij}(r, t)$ is a Green function for thermal stress tensor component σ_{ij}, $ij = r,\theta,z$.

For simple geometries like a plate, cylinder or sphere, Green's functions for temperature and stress can be determined analytically by solving one-dimensional transient heat conduction problem. In case of more complicated shapes, it is necessary to use numerical methods, for example, finite element method in order to obtain accurate Green's functions.

In order to solve Duhamel's integral given by Eq. (11), it is transformed from a continuous into a discrete form:

$$\sigma_{ij}(r,t) = G_{s,ij}(r)T_s(t) + \sum_{t-t_d}^{t} G_{t,ij}(r,t-\tau)\Delta T_s(\tau) \tag{12}$$

where t_d is a cutoff time, $G_{s,ij}(r)$ is a value of the influence function in steady state, while $G_{t,ij}(r)$ is a transient part of the influence function.

Green's functions are determined individually for each supervised region, and in case of thermal stresses, these functions must be determined for each component of the stress tensor σ_{ij}.

An example of Green's function for temperature and stress components at a step increase of fluid temperature by 1°C heating the external surface of a long cylinder is shown in **Figure 1**. The calculations were performed with a constant heat transfer coefficient α = 1000 W/m^2/K and constant material physical properties evaluated at temperature 350°C. The temperature (**Figure 1a**) and stress (**Figure 1b**) responses are presented for the cylinder axis and its outer surface heated by steam.

Green's function for surface temperature *X_surf* initially rapidly increases and reaches nearly 1 after approx. 10^4 s. The temperature at cylinder axis *X_cent* shows some inertia, and a delay time of temperature response equal to 500 s is seen. After this time, the temperature increases at approximately constant rate, and after c.a. 4000 s, the rate of variation visibly starts to decrease. The cylinder surface response is typical for surface-type sensor, while Green's function at cylinder axis is a response of middle-type sensor. The stress response at both areas is a result of temperature variations: axial and circumferential stresses at cylinder surface are equal to each other and rapidly grow reaching a minimum (compressive stress) and subsequently tend slowly to zero; the stresses at cylinder axis grow more slowly,

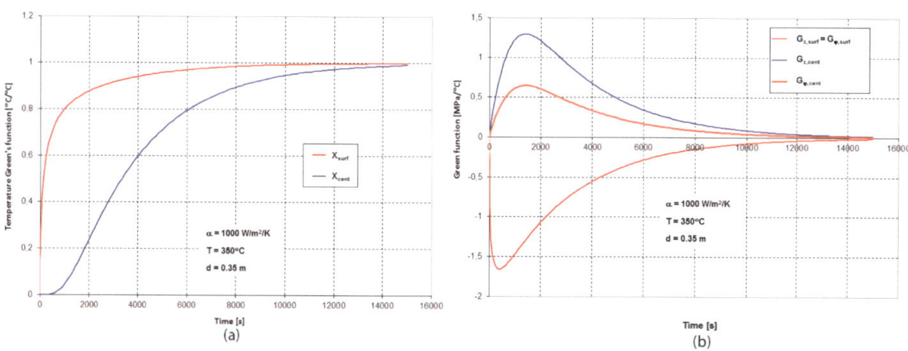

Figure 1.
Green's functions for temperature X *(a) and stress* G *(b) for externally heated cylinder.*

their maxima are lower than at surface (axial stress is two times higher than circumferential stress) and are tensile stresses (sign +). All stress components within the cylinder tend to zero and a stress-free state is reached after approx. 1.4×10^4 s, which corresponds to the instant when temperature within the whole cylinder is uniform.

4. Green's function for variable physical properties and heat transfer coefficient

The above-presented Green functions were determined with constant physical properties and constant heat transfer coefficient. The physical properties of steel influencing temperature (specific heat c_p, thermal conductivity λ) and stress (Young's modulus E, Poisson's number ν, thermal expansion coefficient β) distribution strongly depend on temperature, while the heat transfer coefficient varies with temperature and flow velocity (via the Reynolds number Re). Among the considered physical properties, the most varying is the specific heat which, for a typical low-alloy heat-resistant steel, increases by more than 50% in the temperature range of 20–500°C. The largest drop is observed for the Young modulus which decreases by more than 30% in the same temperature range. Heat transfer coefficient depends mostly on the flow velocity and during turbine start-up can change even by two orders of magnitude.

A consequence of the described variations of physical properties and heat transfer coefficient is a dependence of the Green function on these parameters. The dependence is illustrated in **Figure 2** showing the Green function variations for axial stress at cylinder surface calculated by the finite element method at different temperatures and heat transfer coefficients. Each individual curve corresponds to stress variation determined at constant temperature $T =$ const and constant heat transfer coefficient $\alpha =$ const. As it is seen from the plots, the largest effect on the Green function is observed for the heat transfer coefficient, which significantly affects the stress maxima, time of their occurrence and decay rate. The influence of temperature variation is much smaller and most visible from the instant of maximum stress to the moment of their vanishing.

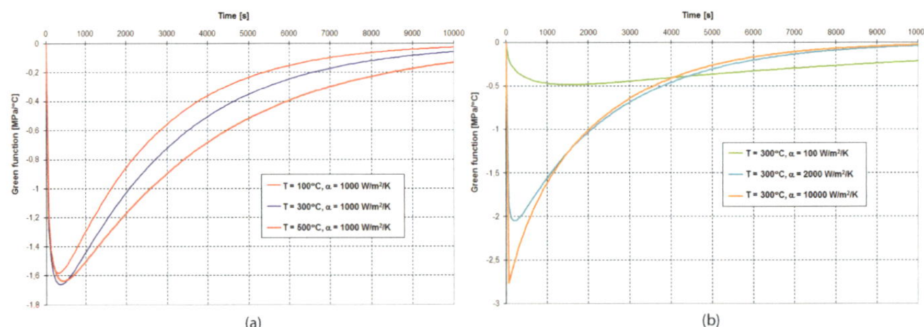

Figure 2.
Green's functions for axial stress at cylinder surface for various temperatures and constant heat transfer coefficient (a) and for constant temperature and various heat transfer coefficients (b).

5. Practical methods for thermoelastic stress calculation

In order to consider the variation of the Green function in Eq. (11), two methods are proposed:

- Use of an equivalent steam temperature determined with a constant heat transfer coefficient $\alpha = \text{const} = \alpha_c$
- Use of an equivalent Green's function determined with variable heat transfer coefficient and physical properties

5.1 Equivalent steam temperature T_{eq}

A flow chart of the proposed stress calculation algorithm using an equivalent steam temperature is presented in **Figure 3** [21]. The calculation process is split into two steps: thermal analysis (step 1) and stress analysis (step 2). In step 1, thermal calculations are performed online, and one-dimensional (1D) heat conduction problem with time-dependent boundary conditions and temperature-dependent physical properties is solved using the finite difference method (FDM). Boundary conditions are given by the measured steam temperature and heat transfer coefficient computed on the basis of the steam mass flow rate. By solving the 1D heat conduction equation, one obtains the radial temperature distribution in the component model. The use of FDM enables full consideration of nonlinearities introduced by time-dependent boundary conditions and temperature-dependent physical properties. Knowing the thermal boundary conditions and radial temperature distribution $T(r,t)$, the equivalent steam temperature $T_{eq}(t)$ for the constant heat transfer coefficient α_c is computed from the surface heat flux (see Eq. (22)). Calculations of the equivalent steam temperature are performed online at each time step for the same constant heat transfer coefficient α_c.

In step 2, thermal stress calculations are performed online with the help of the Duhamel integral. Input for the integral is given by the equivalent steam temperature from step 1 and the Green functions determined offline by means of two/three dimensional (2D/3D) finite element method (FEM) calculations carried out for the real component geometry and constant heat transfer coefficient α_c.

Heat conduction in a homogeneous isotropic solid is described by the Fourier-Kirchhoff differential equation [22]

$$\text{div}[\lambda \text{grad} T(x,t)] + g(x,t) = \rho c_p \frac{\partial T(x,t)}{\partial t} \tag{13}$$

where T is the metal temperature, g is the heat source and c_p is the specific heat.

Neglecting the heat source g and assuming heat conduction in the radial direction r only, Eq. (13) is re-written in a cylindrical coordinate system as follows:

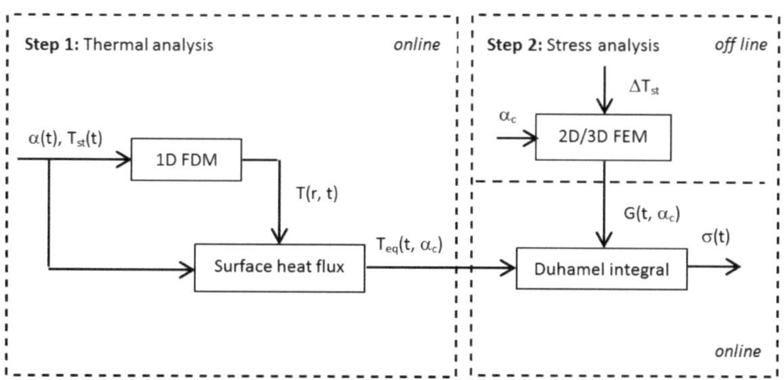

Figure 3.
Flow chart of the stress calculation algorithm with equivalent steam temperature.

$$\rho c_p \frac{\partial T}{\partial t} = \frac{1}{r}\frac{\partial}{\partial r}\left(\lambda r \frac{\partial T}{\partial r}\right) \tag{14}$$

A uniform temperature distribution $T(r) = T_0$ is assumed as initial condition at $t = 0$. Boundary conditions are defined depending on the component modeled. For rotor sections modeled as an externally heated solid cylinder, the boundary conditions are.

$$\left.\frac{\partial T}{\partial r}\right|_{r=0} = 0, \text{ for } t \geq 0 \tag{15}$$

$$\lambda \left.\frac{\partial T}{\partial r}\right|_{r=r_{out}} = -\alpha\left(T_{surf} - T_{st}\right), \text{ for } t \geq 0 \tag{16}$$

where T_{surf} is the metal surface temperature.

For a hollow cylinder model, condition (16) directly applies to the outer surface, and for the inner surface, the boundary condition assumes the form.

$$\left.\frac{\partial T}{\partial r}\right|_{r=r_{in}} = 0, \text{ for } t \geq 0 \tag{17}$$

In case of turbine and valve casings, their thermal behavior is approximated by an internally heated hollow cylinder with the following boundary conditions:

$$\lambda \left.\frac{\partial T}{\partial r}\right|_{r=r_{in}} = -\alpha\left(T_{surf} - T_{st}\right), \text{ for } t \geq 0 \tag{18}$$

$$\left.\frac{\partial T}{\partial r}\right|_{r=r_{out}} = 0, \text{ for } t \geq 0 \tag{19}$$

The one-dimensional heat conduction Eq. (14) is numerically solved using the finite difference method [23]. The material physical properties are temperature-dependent, and their variation is described by polynomial functions. The variation of heat transfer coefficient $\alpha(t)$ is considered by the Nusselt number Nu changing in time as a function of the Reynolds number Re and Prandt number Pr:

$$\mathrm{Nu} = \mathrm{f}(\mathrm{Re}, \mathrm{Pr}) \tag{20}$$

The Nusselt number is defined as follows:

$$\mathrm{Nu} = \frac{\alpha d}{\lambda} \tag{21}$$

where d denotes the cylinder characteristic diameter. A detailed form of Eq. (20) depends on the component geometry and flow character [24]. Practical methods of determining heat transfer coefficients are presented in [3]. The heat transfer model adopted in this study is based on the well-proven formulae for heat transfer coefficients in steam turbine components and can be employed in online calculations of temperatures and thermal stresses which are not possible when more advanced thermal FSI (fluid–structure interaction) modeling is adopted [25–28].

The convective surface heat flux defined by Eq. (16) or (18) with time-dependent heat transfer coefficient $\alpha(t)$ can be alternatively calculated using an equivalent steam temperature T_{eq} and a constant heat transfer coefficient α_c [21]:

$$\alpha\left(T_{surf} - T_{st}\right) = \alpha_c\left(T_{surf} - T_{eq}\right) \tag{22}$$

This relationship can be used to determine the equivalent steam temperature T_{eq} which together with the assumed constant value of heat transfer coefficient α_c ensures the same surface heat flux. The equivalent steam temperature evaluated as

$$T_{eq} = T_{surf} - \frac{\alpha}{\alpha_c}\left(T_{surf} - T_{st}\right) \tag{23}$$

is transferred to step 2 (**Figure 3**) for stress calculations.

The algorithm basing on the equivalent steam temperature has a capability of modeling different heat transfer mechanisms. The conventional application of the Duhamel integral is based on predefined Green functions evaluated for a constant heat transfer coefficient assuming convection heat transfer mode, which is not the only heat exchange mechanism occurring in steam turbines operation. When the turbine starts from a cold state, steam condensation can take place on the rotor outer surfaces or casing inner surfaces, and condensation heat transfer should be taken into account. It is assumed that condensation occurs on the wall, if the following conditions are satisfied [21]:

- The steam temperature is below the saturation temperature at the local pressure p

$$T_{st} < T_{sat}(p) \tag{24}$$

- The heat flux from the water film to metal is greater than the heat flux from steam to the water film

$$\alpha_{cond}\left(T_{sat} - T_{surf}\right) > \alpha_{conv}\left(T_{st} - T_{sat}\right) \tag{25}$$

Condensation heat transfer takes place with very high heat transfer coefficients resulting in high heat fluxes on the component surface. This, in turn, brings about high surface temperature rates producing elevated thermal stresses on the surface or in the stress concentration areas. The most favorable conditions for the condensation to take place prevail at the initial phase of cold start-ups when the metal temperature is low, that is, below 100°C, and the steam saturation temperature high enough to produce high heat fluxes. This phenomenon is minimized by prewarming steam turbine components and supplying steam of possibly highest superheating.

5.2 Equivalent Green's function

The second method relies on using an equivalent Green's function in Eq. (11) determined, assuming a variable heat transfer coefficient and material physical properties. It can be done if the variation of process parameters is known in advance and used to determine the shape of the equivalent Green's function. This is the case in steam turbines operation as steam parameters are defined in design start-up diagrams for specific types of starts and should be followed within some margins in real operation. The method assumes little deviations of real operating conditions from those defined by the start-up diagrams.

The equivalent Green function is evaluated by performing finite element thermal analysis with the following boundary conditions [29]:

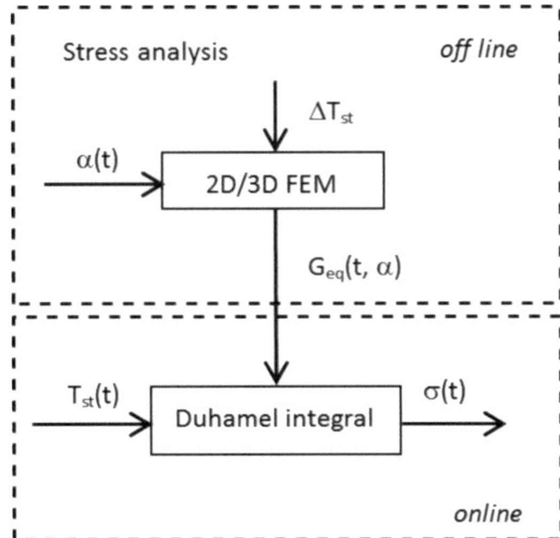

Figure 4.
Flow chart of the stress calculation algorithm with equivalent Green function.

1. Steam temperature step characteristic for a given start-up type.

2. Variable heat transfer coefficient resulting from the variation in time of the steam mass flow rate characteristic for a given start-up.

The flow chart of the thermoelastic stress calculation algorithm based on the equivalent Green function is shown in **Figure 4**. The equivalent Green function is pre-computed offline by performing finite element calculations in 2D or 3D model of the real component. The finite element model is nonlinear due to the temperature-dependent material properties and time-dependent heat transfer coefficient. The equivalent Green functions are evaluated for each stress component and for various transient events like a cold, warm and hot start-up and shutdown. These events are characterized by specific variations of the heat transfer coefficients and the range of temperature variations affecting the material physical properties. The appropriate Green function is selected at the beginning of the transient event and used to calculate the thermal stress online with the help of the Duhamel integral.

6. Validation of the methods for turbine valve

First validations of both methods for a simple cylinder heated externally by steam were presented in [21] for the equivalent steam temperature and in [29] for the equivalent Green's function. A more complex example of a steam turbine valve operating at highly variable steam conditions, like temperature, pressure and flow rate, was used in this study to compare and validate the proposed algorithms. Numerical calculations were performed using a FEM model, and the presented algorithms with transient boundary conditions corresponding to a cold start-up are shown in **Figure 5**. The equivalent steam temperature was obtained for the calculated surface temperature and the assumed heat transfer coefficient $\alpha_c = 500$ W/m^2 K^{-1}. The equivalent Green functions were computed assuming the variation of heat transfer coefficient shown in **Figure 5**. As it is seen, the equivalent steam temperature exceeds the real steam temperature at a time point where the actual heat transfer coefficient rises above the assumed constant value of α_c, which is a consequence of the equal surface heat fluxes assumed.

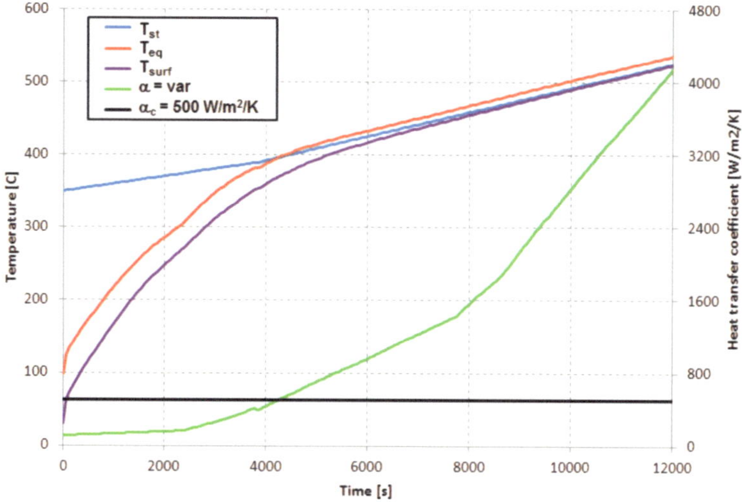

Figure 5.
Thermal boundary conditions for cold start.

Figure 6 presents the variation of surface heat flux $Q_{surf} = \alpha(T_{surf} - T_{st})$ and surface temperature rate dT_{surf}/dt during the cold start-up. Both curves exhibit two local minima: the first one after 2340 s and the second one after 3840 s. They correspond to the instants of the change in slope of the heat transfer coefficient curve shown in **Figure 5** and will affect the shape of the Green functions and the variation of the equivalent Huber-Mises stress.

The calculations were performed with variable material physical properties depending on metal temperature. **Figure 7** presents their variation during the cold start-up at the point of highest thermoelastic stresses located at the bowl inner surface. The largest drop is observed for the Young modulus which changes by around 30% during the cold start, while the largest rise occurs for the specific heat varying by more than 50% at the same time.

Figures 8 and **9** show the temperature and Huber-Mises stress distributions in the valve casing at three instants of the cold start calculated with the FEM model. These stresses were used to validate the proposed algorithms by comparing them

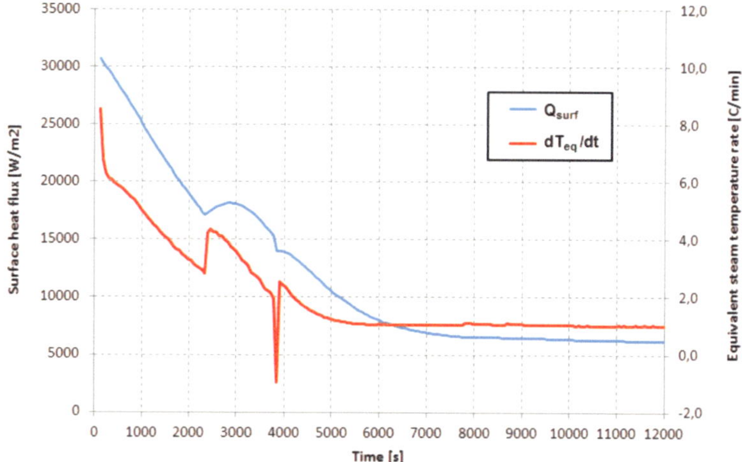

Figure 6.
Variation of surface heat flux and equivalent steam temperature rate during the cold start.

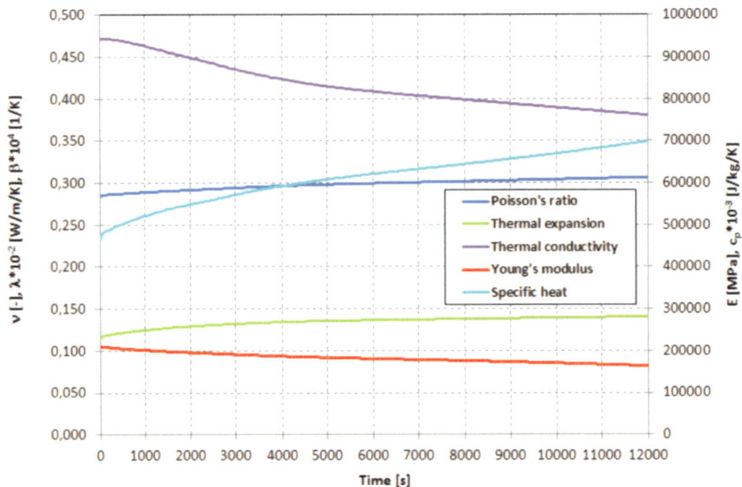

Figure 7.
Variation of physical properties during the cold start.

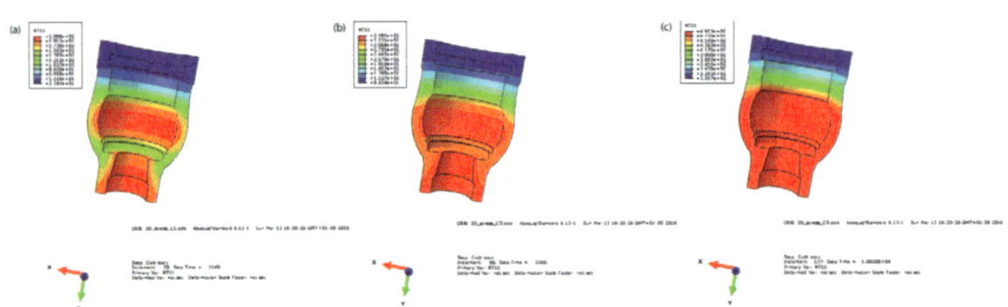

Figure 8.
Instantaneous temperature distribution during the cold start after 1140 s (a), 3360 s (b) and 10,020 s (c).

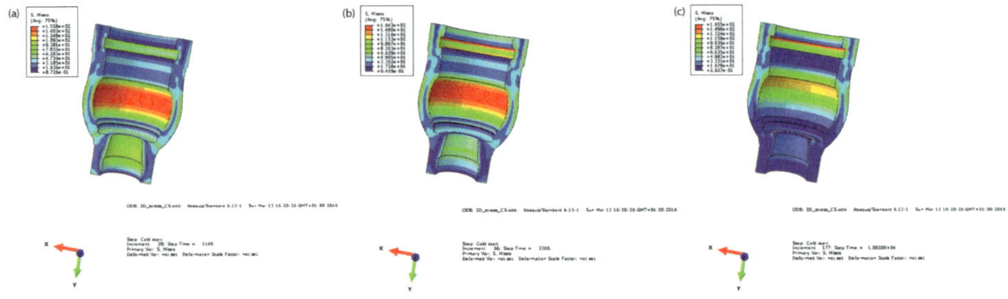

Figure 9.
Instantaneous Huber-Mises stress distribution during the cold start after 1140 s (a), 3360 s (b) and 10,020 s (c).

with the predictions of both methods. The temperature field in the valve casing is two-dimensional, which results from both the non-regular geometry and non-uniform heat transfer conditions prevailing on the heated surfaces. The most intensive heat transfer takes place in the bowl and diffuser seat area, and the inner surface temperature is highest in these regions. The highest stress occurs at the inner surface of the valve bowl where both the temperature and stress gradients are predominantly present in the radial direction. Thermal stress distribution in the bowl wall (**Figure 9**) is similar to the typical distribution in a cylindrical or a spherical wall with stress maxima occurring on the outer and inner surfaces and the minimum stress present approximately in the middle of the wall thickness. The FEM model was also used for determining the equivalent Green functions and Green functions for a constant heat transfer coefficient $\alpha_c = 500$ W/m^2/K in the bowl region and variable physical properties corresponding to the temperature range 30–525°C. The Green functions for each stress component at the bowl inner surface for the constant heat transfer coefficient are shown in **Figure 10** and for the varying heat transfer coefficient are presented in **Figure 11**. All stresses are compressive, and except for the radial component, rapidly increase to their maxima and then slowly decay. The largest stresses are generated in axial and circumferential directions (G_{22} and G_{33}, respectively), perpendicular to the direction of the highest temperature gradient. These Green functions implicitly include also the effect of 2D temperature distribution resulting from different heat transfer conditions

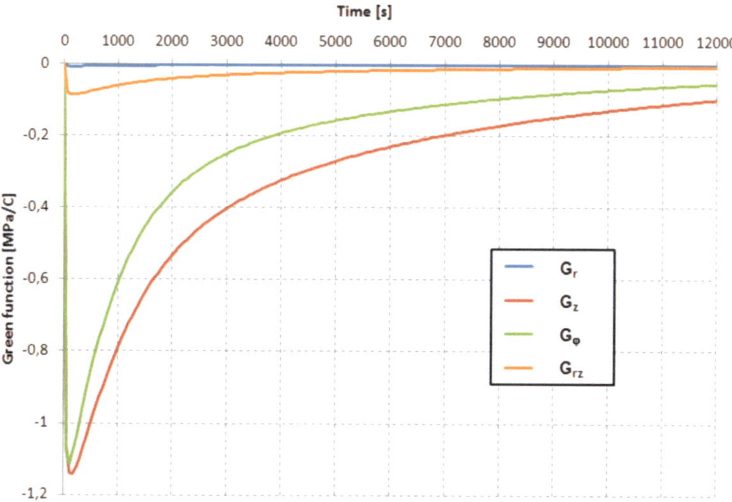

Figure 10.
The Green functions for stress components at bowl maximum stress location for constant heat transfer coefficient.

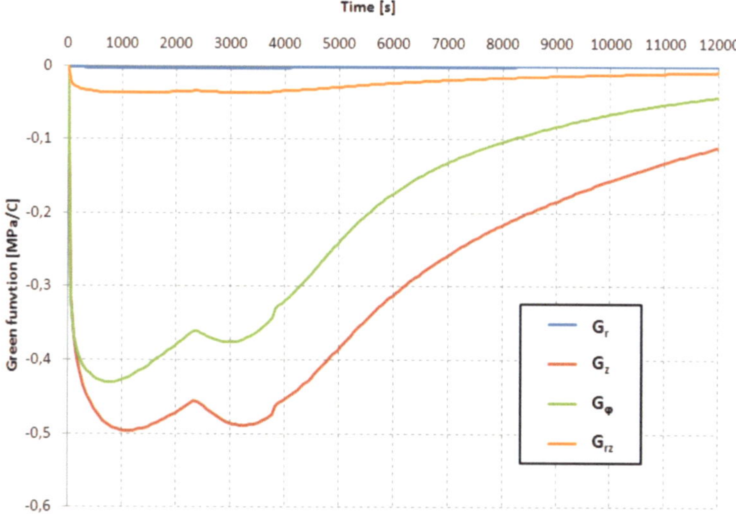

Figure 11.
The Green functions for stress components at bowl maximum stress location for varying heat transfer coefficient.

prevailing on different valve surfaces heated by steam. This effect was accounted for by applying to these surfaces the heat transfer coefficient values different than α_c = 500 W/m^2/K (in the equivalent steam temperature method) or α = var (in the equivalent Green function method) which was achieved by applying the appropriate multiplication factors to the base heat transfer coefficients. There is a clear qualitative difference observed between the Green functions obtained with the constant and variable heat transfer coefficient. When the applied heat transfer coefficient is constant during the heating process, all the Green functions exhibit one maximum at the very beginning of the process, followed by continuous stress decay (**Figure 10**). The effect of non-monotonic variation of the surface heat flux is not included here in the Green function but is taken into account in the equivalent steam temperature rate exhibiting two local minima (**Figure 6**). When the applied heat transfer coefficient is varying during the heating process, all the Green functions exhibit two maxima: the first one at the beginning of the process and resulting from the initial temperature step and the second one corresponding to the increase of the surface heat flux (**Figure 11**).

The variation of Huber-Mises stress at the bowl obtained from the FEM analysis is compared in **Figure 12** with the predictions of both simplified algorithms. Calculations were carried out using the Green functions shown in **Figures 10** and **11**, and the steam temperatures presented in **Figure 5**. As it is seen, both methods predict the occurrence of two stress maxima with high accuracy as compared with the FE results. A slightly better agreement is found for the method using the equivalent Green function which underpredicts the first maximum by 1.1% and the second maximum by 2.5%. The instants of both maxima correspond to those obtained from the FE model. The method using the equivalent steam temperature overpredicts both the first and second stress maximum with an error of 2.8 and 3.3%, respectively. A worse agreement with the FE results is also observed for the instants of stress maximum: the first maximum occurs later while the second one sooner as compared with the FE calculation results.

The occurrence of two subsequent stress maxima predicted in the finite element analysis is a result of variable surface heat flux exhibiting in time local maxima and minima. This effect was reproduced by both simplified methods but in a different way. In the equivalent Green function method, this function had two maxima and in combination with the applied steam temperature profile resulted in the two stress

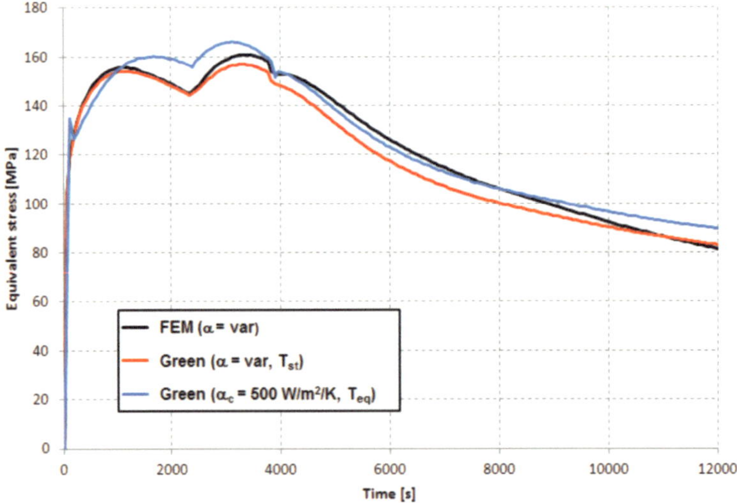

Figure 12.
Equivalent stress variations during the cold start-up.

maxima. In the equivalent steam temperature method, this effect was obtained as a result of specific variation of the equivalent steam temperature applied together with the Green function of a typical shape.

Also, the stress decay period persisting after 4000 s of the cold start-up is accurately modeled by the two simplified methods. The relative error in this phase is generally below 10%, and initially, both methods underpredict the FE equivalent stress with a tendency to overpredict it at longer times.

7. Summary

The presented methods for online calculation of the thermoelastic stresses in steam turbine components are based on the Green function and Duhamel integral and consider the effect of variable heat transfer coefficient and material physical properties on thermal stresses. This effect is taken into account either by using the equivalent steam temperature determined with a constant heat transfer coefficient or by applying the equivalent Green's function determined with variable heat transfer coefficient and physical properties.

The first method is based on a numerical solution of 1D heat conduction problem and the Duhamel integral for thermal stress calculation. The equivalent steam temperature is calculated in step 1 and then used as an input to thermal stress model employing the Green function evaluated with a constant heat transfer coefficient. This approach allows for a full consideration of nonlinearities introduced by variable physical properties and heat transfer coefficient in temperature calculations. The effect of variable heat transfer coefficient significantly affecting the thermal stress results was taken into account by introducing the equivalent steam temperature evaluated on the basis of surface heat flux. The minor influence of variable physical properties on the stress response was considered by determining the Green functions from a step change of steam temperature in the range of its possible variation.

The second method relies on using the equivalent Green's function determined assuming a variable heat transfer coefficient and material physical properties. The equivalent Green function is evaluated by performing finite element thermal analysis for a step change of steam temperature and variable heat transfer coefficient resulting from the variation in time of the steam mass flow rate.

The effectiveness of both methods was shown by comparing their predictions with the results of finite element calculations of a steam turbine valve. The multidimensional geometry and non-uniform thermal boundary conditions result in two-dimensional temperature and stress fields calculated using the FEM model. The comparison of equivalent stresses in the casing bowl calculated using the FEM model and the proposed methods revealed their good accuracy. The relative errors of maximum stress prediction are in the range of 1–3% and can be considered as very low taking into account the complex geometry of the valve, nonlinear boundary conditions and a broad range of temperature variation inducing significant variation of material physical properties. The methods can be applied for online calculation of thermoelastic stress utilized in thermal stress control and lifetime monitoring systems of steam turbines.

Nomenclature

c_p [J/kg/K]	specific heat
d [m]	characteristic diameter
e [−]	volumetric strain

E [MPa]	Young modulus
g [W/m^3]	heat source
G [MPa]	shear modulus
G_{ij} [MPa/°C]	Green's function tensor
G_{eq} [MPa/°C]	equivalent Green's function
G_s [MPa/°C]	steady-state Green's function
G_t [MPa/°C]	transient Green's function
H [°C]	Heaviside's function
n [—]	normal direction
Nu [—]	Nusselt number
p [MPa]	pressure
p_b [MPa]	pressure due to centrifugal forces of blades
Pr [—]	Prandtl number
Q_{surf} [W/m^2]	surface heat flux
r_{in} [m]	inner radius
r_{out} [m]	outer radius
Re [—]	Reynolds number
t [s]	time
t_d [s]	cutoff time
T [°C]	temperature
T_{eq} [°C]	equivalent steam temperature
T_{sat} [°C]	saturation temperature
T_{st} [°C]	steam temperature
T_{surf} [°C]	surface temperature
T_0 [°C]	initial temperature
u [m]	radial displacement
w [m]	axial displacement
X [°C/°C]	Green's function for temperature
z [m]	axial coordinate
α [W/m^2/K]	heat transfer coefficient
α_{cond} [W/m^2/K]	condensation heat transfer coefficient
α_{conv} [W/m^2/K]	convection heat transfer coefficient
α_c [W/m^2/K]	constant heat transfer coefficient
β [1/K]	thermal expansion coefficient
γ_{rz} [—]	shear strain component
Δ [—]	difference
ε_r [—]	radial strain component
ε_φ [—]	circumferential strain component
ε_z [—]	axial strain component
λ [W/m/K]	thermal conductivity
ν [—]	Poisson's ratio
ρ [kg/m^3]	density
σ_n [MPa]	normal stress
σ_r [MPa]	radial stress component
σ_φ [MPa]	circumferential stress component
σ_z [MPa]	axial stress component
σ_{rz} [MPa]	shear stress component
σ_{ij} [MPa]	stress tensor
τ [s]	artificial time
ω [rad/s]	angular velocity
∇^2[—]	Laplacian

Author details

Mariusz Banaszkiewicz* and Janusz Badur
The Szewalski Institute of Fluid-Flow Machinery Polish Academy of Sciences, Gdańsk, Poland

*Address all correspondence to: mbanaszkiewicz@imp.gda.pl

References

[1] Vogt J, Schaaf T, Helbig K. Optimizing lifetime consumption and increasing flexibility using enhanced lifetime assessment methods with automated stress calculation from long-term operation data. In: Proceedings of the ASME Turbo Expo; 3-7 June 2013; San Antonio

[2] Rusin A, Łukowicz H, Lipka M, Banaszkiewicz M, Radulski W. Continuous control and optimisation of thermal stresses in the process of turbine start-up. In: 6th International Congress on Thermal Stresses; 26-29 May 2005; Vienna

[3] Taler J. Theory and Practice of Heat Transfer Processes Identification. Zakład Narodowy im. Ossolińskich: Wrocław; 1995 (in Polish)

[4] Rusin A, Nowak G, Lipka M. Practical algorithms for online stress calculations and heating process control. Journal of Thermal Stresses. 2014; **37**(11):1286-1301

[5] Rusin A, Radulski W, Banaszkiewicz M. The Use of Duhamel's Integral in Lifetime Supervision Systems of Steam Turbines. In: 5th Conference on Research Problems in Power Engineering. 4-7 December 2001; Warsaw

[6] Rusin A, Łukowicz H, Malec A, Banaszkiewicz M, Lipka M. Optimisation of start-up parameters of steam turbines. In: 6th Conference on Research Problems in Power Engineering. 9-12 December 2003; Warsaw

[7] Taler J, Węglowski B, Zima W, Duda P, Grądziel S, Sobota T, Cebula A, Taler D. Computer system for monitoring power boiler operation. Proceedings of IMechE, Part A: Journal of Power and Energy. 2008;**222**:13-24

[8] Lee HY, Kim JB, Yoo B. Green's function approach for crack propagation problem subjected to high cycle thermal fatigue loading. International Journal of Pressure Vessels and Piping. 1999;**76**: 487-494

[9] Song G, Kim B, Chang S. Fatigue life evaluation for turbine rotor using Green's function. Proceedia Engineering. 2011;**10**:2292-2297

[10] Koo GH, Kwon JJ, Kim W. Green's function method with consideration of temperature dependent material properties for fatigue monitoring of nuclear power plants. International Journal of Pressure Vessels and Piping. 2009;**86**:187-195

[11] Botto D, Zucca S, Gola MM. A methodology for on-line calculation of temperature and thermal stress under non-linear boundary conditions. International Journal of Pressure Vessels and Piping. 2003;**80**:21-29

[12] Zhang HL, Liu S, Xie D, Xiong Y, Yu Y, Zhou Y, Guo R. Online fatigue monitoring models with consideration of temperature dependent properties and varying heat transfer coefficients. Science and Technology of Nuclear Installations. Vol. 2013. Article ID 763175

[13] Adamowicz A. Axisymmetric FE model to analysis of thermal stresses in a brake disc. Journal of Theoretical and Applied Mechanics. 2015;**53**(2):357-370

[14] Adamowicz A. Finite element analysis of the 3D thermal stress state in a brake disc. Journal of Theoretical and Applied Mechanics. 2016;**54**(1):205-218

[15] Zienkiewicz OC, Taylor RL, Zhu JZ. The Finite Element Method: It's Basis and Fundamentals. Burlington: Elsevier Butterworth-Heinemann; 2005; 733 p

[16] Orłoś Z. Thermal Stresses. Warsaw: PWN; 1991 (in Polish)

[17] Hetnarski RB, Eslami MR. Thermal Stresses - Advanced Theory and Applications. Dordrecht, Netherlands: Springer; 2009. p. 559

[18] Nowacki W. Thermoelasticity. Warsaw: PWN; 1986 (in Polish)

[19] Badur J, Nastałek L. Thermodynamics of thermo-deformable bodies. In: Hetnarski RB, editor. Encyclopedia of Thermal Stresses. Dordrecht, Netherlands: Springer; 2014

[20] Carslow HS, Jaeger JC. Conduction of Heat in Solids. Oxford, UK: Oxford University Press; 1959

[21] Banaszkiewicz M. Online determination of transient thermal stresses in critical steam turbine components using a two-step algorithm. Journal of Thermal Stresses. 2017;**40**: 690-703

[22] Noda N, Hetnarski RB, Tanigawa Y. Thermal stresses. Abingdon, UK: Taylor & Francis; 2003

[23] Hetnarski RB, editor. Encyclopedia of Thermal Stresses. Dordrecht, Netherlands: Springer; 2014

[24] Chmielniak T, Kosman T. Thermal Loads of Steam Turbines. Warsaw: WNT; 1990 (in Polish)

[25] Badur J, Ziółkowski P, Kornet S, Stajnke M, Bryk M, Banaś K, Ziółkowski PJ. The effort of the steam turbine caused by a flood wave load. In: AIP Conference Proceedings; 2017

[26] Badur J, Ziółkowski P, Sławiński D, Kornet S. An approach for estimation of Water Wall degradation within pulverized-coal boilers. Energy. 2015; **92**:142-152

[27] Badur J, Ziółkowski P, Zakrzewski W, Sławiński D, Kornet S, Kowalczyk T, Hernet J, Piotrowski R, Felicjancik J, Ziółkowski PJ. An advanced thermal-FSI approach to flow heating/cooling. Journal of Physics: Conference Series; **530**(2015):012039

[28] Banaś K, Badur J. Influence of strength differential effect on material effort of a turbine guide vane based on a Thermoelastoplastic analysis. Journal of Thermal Stresses. 2017;**40**(11): 1368-1385

[29] Banaszkiewicz M. Online monitoring and control of thermal stresses in steam turbine rotors. Applied Thermal Engineering. 2016;**94**:763-776